Annals of Mathematics Studies

Number 95

C^*-ALGEBRA EXTENSIONS AND K-HOMOLOGY

BY

RONALD G. DOUGLAS

PRINCETON UNIVERSITY PRESS
AND
UNIVERSITY OF TOKYO PRESS

PRINCETON, NEW JERSEY
1980

Published in Japan exclusively by
University of Tokyo Press
In other parts of the world by
Princeton University Press

Printed in the United States of America
by Princeton University Press, Princeton, New Jersey

Library of Congress Cataloging in Publication data will
be found on the last printed page of this book

CONTENTS

PREFACE

In this book I have written up the Hermann Weyl Lectures, which I gave at the Institute for Advanced Study during February, 1978. My contribution to the work on which I reported was done in collaboration with L. G. Brown of Purdue University and P. A. Fillmore of Dalhousie University. The basic references are [20], [21], [22]. I will not repeat all the references given there although I will give the more important ones and recent papers will be cited in more detail. As we indicated in [20] and [22], there are a number of people to whom we are indebted. I cannot mention them all but would like to acknowledge the influence of M. F. Atiyah and I. M. Singer. In addition, I would like to thank Jerry Kaminker and Claude Schochet for discussions on this material, especially in connection with Chapters five and six. Finally, I would like to express my appreciation to my audience for their interest which spurred me to make this exposition more comprehensive than I had originally planned.

R. G. DOUGLAS

C*-Algebra Extensions
and K-Homology

CHAPTER 1
AN OVERVIEW

Although there are no doubt many possible connections between operator theory and algebraic topology, in this book I concentrate on one interplay between the two subjects. The machinery which establishes this connection solves various problems in operator theory which will presently be described and suggests many others. Moreover, exciting applications in algebraic topology seem within reach. We shall say something about these a little later. In this chapter we want to give an overview of our topic including the origins and a general outline of the theory.

Quite appropriately this work can be traced back to a theorem of Hermann Weyl concerning the behavior of the spectrum of a formally self-adjoint differential operator under a change of boundary conditions. A converse due to von Neumann and the evolution of the abstract theories of Fredholm operators and of operator algebras were necessary steps for its development. More recently, the connection between Fredholm operators, index theory, and K-theory developed by Atiyah, Singer, Janich and others set the stage. Finally, the general interest of operator theorists in problems involving compact perturbations provided the particular impetus to this work. These things will be discussed in more detail after the abstract notion which lies at the center of this work is introduced.

We shall study a certain class of C^*-extensions

$$0 \to \mathcal{K}(\mathfrak{H}) \to \mathcal{E} \to C(X) \to 0$$

of the C^*-algebra $\mathcal{K}(\mathfrak{H})$ of the compact operators by the C^*-algebra $C(X)$ of continuous complex-valued functions on a compact metrizable space X. Recall that an algebra $\mathcal{E}$ over $\mathbb{C}$ is said to be a C^*-*algebra* if $\mathfrak{E}$ possesses a norm $\|\cdot\|$ relative to which $\mathfrak{E}$ is a Banach algebra and an involution $a \to a^*$ satisfying $\|a^*a\| = \|a^*\|\,\|a\|$ for a in $\mathfrak{E}$. If $\mathfrak{H}$ is a complex Hilbert space and $\mathfrak{L}(\mathfrak{H})$ denotes the algebra of bounded linear operators on $\mathfrak{H}$ then $\mathfrak{L}(\mathfrak{H})$ is a C^*-algebra with the operator norm and adjoint as involution. More generally, closed, self-adjoint subalgebras of $\mathfrak{L}(\mathfrak{H})$ are C^*-algebras and a theorem of Gelfand and Naimark asserts that all C^*-algebras up to *-isometrical isomorphism are obtained in this manner. In particular, the algebra $\mathcal{K}(\mathfrak{H})$ of compact operators on $\mathfrak{H}$ is probably the most elementary infinite dimensional C^*-algebra. (Recall that T in $\mathfrak{L}(\mathfrak{H})$ is *compact* if the image of the unit ball under T is compact or, equivalently for operators on a Hilbert space, if T is the norm limit of finite rank operators.) Another simple class of C^*-algebras consists of those which are commutative and another theorem of Gelfand and Naimark states that these are all *-isometrically isomorphic to $C(X)$ for some compact Hausdorff space X. Thus in considering the class of extensions described above we are following a well-established algebraic dictum; that is, study algebras which are obtained as extensions of simpler classes of algebras. However, this was not our original motivation. To explain that let us consider some examples of naturally occurring extensions.

Let T denote the unit circle in $\mathbb{C}$, $L^2(T)$ the Lebesgue space relative to normalized Lebesgue measure, and $H^2(T)$ the Hardy space obtained as the closure of the analytic polynomials in $L^2(T)$ or equivalently, the closed linear span of $\{z^n : n \geq 0\}$. If P denotes the orthogonal projection from $L^2(T)$ onto $H^2(T)$, then for ψ in $C(T)$ the Toeplitz operator T_ψ on $H^2(T)$ is defined by $T_\psi f = P(\psi f)$ for f in $H^2(T)$. If $\mathfrak{T}$ denotes the C^*-subalgebra of $\mathfrak{L}(H^2(T))$ generated by $\{T_\psi : \psi \in C(T)\}$ then Coburn showed in [28] that

$$\mathfrak{T} = \{T_\psi + K : \psi \in C(T), K \in \mathcal{K}(H^2(T))\}$$

and observed that $\mathfrak{T}$ is an extension of $\mathcal{K}$ by $C(T)$. That is, there is a short exact sequence

$$0 \longrightarrow \mathcal{K}(H^2(T)) \xrightarrow{i} \mathcal{T} \xrightarrow{\phi} C(T) \longrightarrow 0\,,$$

where i is inclusion and ϕ is the symbol map defined by $\phi(T_\psi + K) = \psi$. It can be shown (cf. [32]) that the basic index results for Toeplitz operators with continuous symbol follow directly from this. Coburn further pointed out that if $\mathcal{S}$ is defined by

$$\mathcal{S} = \{M_\psi + K : \psi \in C(T), K \in \mathcal{K}(L^2(T))\}\,,$$

where M_ψ is the operator defined to be multiplication by ψ on $L^2(T)$, then an extension of $\mathcal{K}$ by $C(T)$ is obtained which is not equivalent to the Toeplitz extension. Thus he raised the problem of determining all extensions of $\mathcal{K}$ by $C(T)$. About the same time Atiyah and Singer arrived at the same problem for reasons which we will explain presently. Before proceeding we enlarge the class of Toeplitz examples.

If Ω is a strongly pseudo-convex domain in $\mathbb{C}^n$, $L^2(\partial\Omega)$ is the Lebesgue space relative to surface measure on $\partial\Omega$, and $H^2(\partial\Omega)$ the Hardy space obtained as the closure in $L^2(\partial\Omega)$ of the functions holomorphic on a neighborhood of the closure of Ω, then the analogue of Toeplitz operators can be defined on $H^2(\partial\Omega)$ (cf. [97], [13]) as the compression T_ψ of a continuous multiplier ψ to $H^2(\partial\Omega)$. (It is not necessary in what follows to know the definition of strongly pseudo-convex but the ball is strongly pseudo-convex while the bidisk is not.) Moreover the sequence

$$0 \longrightarrow \mathcal{K}(H^2(\partial\Omega)) \xrightarrow{i} \mathcal{T}_\Omega \xrightarrow{\phi_\Omega} C(\partial\Omega) \longrightarrow 0$$

can be shown to be exact, where $\mathcal{T}_\Omega$ is the C^*-subalgebra of $\mathcal{L}(H^2(\partial\Omega))$ generated by the Toeplitz operators with continuous symbol, i is inclusion, and ϕ_Ω is the symbol map defined by $\phi_\Omega(T_\psi + K) = \psi$. Thus there is a naturally occurring class of Toeplitz extensions. (Before continuing we point out that a similar construction can be carried out using the Lebesgue space relative to volume measure on Ω. The symbol space is still $C(\partial\Omega)$, however.)

In a different direction let M be a closed differentiable manifold and let $L^2(M)$ denote the Lebesgue space defined relative to a fixed smooth measure on M. Now a zero'th order pseudo-differential operator with scalar coefficients defines a bounded operator on $L^2(M)$ and we let $\mathcal{P}_M$ denote the C^*-subalgebra of $\mathcal{L}(L^2(M))$ generated by all such operators together with $\mathcal{K}(L^2(M))$. (I believe adding $\mathcal{K}(L^2(M))$ as generators is unnecessary if M is connected.) Then we obtain the extension

$$0 \longrightarrow \mathcal{K}(L^2(M)) \xrightarrow{\ i\ } \mathcal{P}_M \xrightarrow{\ \phi_M\ } C(S^*(M)) \longrightarrow 0$$

where i is inclusion, $S^*(M)$ is the cosphere bundle on M, and ϕ_M is the symbol map. Moreover, this extension is intimately related to the Atiyah-Singer index theorem and other results, and it was in this connection that Atiyah and Singer were interested in classifying the extensions of $\mathcal{K}$ by $C(T)$.

One of our main sources of motivation sprang from operator theory. Let T be an *essentially normal operator* on $\mathfrak{H}$, that is, an operator T in $\mathcal{L}(\mathfrak{H})$ such that the self-commutator $[T, T^*] = TT^* - T^*T$ is compact. The problem in operator theory was basically to classify the essentially normal operators up to unitary equivalence modulo $\mathcal{K}$ and to determine the possibilities. Why should one think that such a thing might be possible? To answer that we have to look at the theorem of Weyl to which we alluded earlier. For H a self-adjoint operator, Weyl defined the essential spectrum $\sigma_e(H)$ to be all λ in the spectrum $\sigma(H)$ except for isolated eigenvalues of finite multiplicity. He proved in [101] that if the self-adjoint operators H_1 and H_2 differ by a compact operator, then $\sigma_e(H_1) = \sigma_e(H_2)$. Trivially this implies that if H_1 and H_2 are self-adjoint operators for which there exists a unitary operator U such that

$$U^*H_1U = H_2 + K$$

for some K in $\mathcal{K}(\mathfrak{H})$, then $\sigma_e(H_1) = \sigma_e(H_2)$. About twenty years later von Neumann established [62] the converse, and almost fifty years later in response to a question of Halmos [41] the result was extended to normal operators by Berg [12]. Thus we have

THEOREM 1. *If* N_1 *and* N_2 *are normal operators on* $\mathfrak{H}$, *then the following are equivalent: a) there exists a unitary operator* U *and a compact operator* K *such that* $U^*N_1U = N_2 + K$ *and b)* $\sigma_e(N_1) = \sigma_e(N_2)$.

Moreover, if X *is any compact subset of* $\mathbb{C}$, *then there exists a normal operator* N *such that* $\sigma_e(N) = X$.

This solved completely the problem of classifying normal operators up to unitary equivalence modulo $\mathcal{K}$ and showed that the equivalence classes are in one to one correspondence with compact subsets of $\mathbb{C}$. (This should be compared with the solution to the unitary equivalence problem for normal operators which involves multiplicity theory and hence equivalence classes correspond to cardinal numbers assigned to a Borel partition of the spectrum.) Therefore it seemed reasonable to try to extend this result to essentially normal operators.

Now for an essentially normal operator $T = N + K$ in the algebraic linear span $\mathfrak{N} + \mathcal{K}$, where $\mathfrak{N}$ is the collection of normal operators, the only problem is to obtain $\sigma_e(N)$ in terms of T. That is easy once it is observed that the essential spectrum of a normal operator coincides with the spectrum in the quotient algebra $\mathfrak{L}(\mathfrak{H})/\mathcal{K}(\mathfrak{H})$. More precisely, since $\mathcal{K}(\mathfrak{H})$ is a closed two-sided ideal in $\mathfrak{L}(\mathfrak{H})$ we can define the quotient $\mathfrak{Q}(\mathfrak{H}) = \mathfrak{L}(\mathfrak{H})/\mathcal{K}(\mathfrak{H})$, usually called the *Calkin algebra*, which can be shown to be a C^*-algebra. If $\pi: \mathfrak{L}(\mathfrak{H}) \to \mathfrak{Q}(\mathfrak{H})$ denotes the quotient map, then we can consider the spectrum $\sigma_{\mathfrak{Q}(\mathfrak{H})}(\pi(T))$ for T in $\mathfrak{L}(\mathfrak{H})$ and this spectrum can be shown to coincide with $\sigma_e(T)$ for T normal. Thus the *essential spectrum* for general T is defined to be $\sigma_e(T) = \sigma_{\mathfrak{Q}(\mathfrak{H})}(\pi(T))$. Therefore the Weyl-von Neumann-Berg theorem extends to essentially normal operators T of the form $T = N + K$ since $\sigma_e(N) = \sigma_e(T)$.

Now not all essentially normal operators have this form. The operator T_z on $H^2(T)$ provides such an example but the proof involves the notion of index. Recall that the class of *Fredholm operators* $\mathrm{Fred}(\mathfrak{H})$ can be defined (cf. [32]) such that T is in $\mathrm{Fred}(\mathfrak{H})$ if and only if $\pi(T)$ is invertible in $\mathfrak{Q}(\mathfrak{H})$ if and only 0 is not in $\sigma_e(T)$, or equivalently

$$\mathrm{Fred}(\mathfrak{H}) = \pi^{-1}(\mathfrak{Q}(\mathfrak{H})^{-1}) ,$$

where $\mathfrak{E}^{-1}$ denotes the invertible elements in the algebra $\mathfrak{E}$. Then it follows that

1) $T \in \mathrm{Fred}(\mathfrak{H})$, $K \in \mathfrak{K}(\mathfrak{H})$ implies $T+K \in \mathrm{Fred}(\mathfrak{H})$,
2) $S, T \in \mathrm{Fred}(\mathfrak{H})$ implies $ST \in \mathrm{Fred}(\mathfrak{H})$, and
3) $\mathrm{Fred}(\mathfrak{H})$ is an open subset of $\mathfrak{L}(\mathfrak{H})$.

The latter property follows from the fact that the invertible elements in a Banach algebra form an open set. Moreover, a result of Atkinson shows that T is in $\mathrm{Fred}(\mathfrak{H})$ if and only if i) ran T is closed, ii) $\dim\ker T < \infty$, and iii) $\dim\ker T^* < \infty$. Lastly, the *index* is defined from $\mathrm{Fred}(\mathfrak{H})$ to Z by

$$\mathrm{ind}(T) = \dim\ker T - \dim\ker T^*$$

and is a continuous homomorphism which is invariant under compact perturbation.

The relevance of index to our problem lies in the following. By our earlier remarks $[T_z, T_z^*]$ is compact and $\sigma_e(T_z) = \mathbf{T}$; therefore T_z is Fredholm. Further, relative to the orthonormal basis $\{1, z, z^2, \cdots\}$ for $H^2(\mathbf{T})$, T_z is the unilateral shift (that is, $T_z z^n = z^{n+1}$ for $n \geq 0$), while T_z^* is the backward shift. Thus if $(a_0, a_1 \cdots)$ are the Fourier coefficients for f in $H^2(\mathbf{T})$, then $T_z(a_0, a_1 \cdots) = (0, a_0, a_1 \cdots)$ and $T_z^*(a_0, a_1, \cdots) = (a_1, a_2, \cdots)$. Therefore, $\dim\ker T_z = 0$, and $\dim\ker T_z^* = 1$, and hence $\mathrm{ind}\, T_z = -1$. Since $\|Nx\|^2 = (N^*Nx, x) = (NN^*x, x) = \|N^*x\|^2$ for N normal on $\mathfrak{H}$ and x in $\mathfrak{H}$, it follows that $\ker N = \ker N^*$ and hence $\mathrm{ind}\, N = 0$ if N is Fredholm. Therefore, the assumption that $T_z = N + K$ where N is normal and K is compact leads to the contradiction

$$-1 = \mathrm{ind}(T_z) = \mathrm{ind}(N+K) = \mathrm{ind}(N) = 0 .$$

Now not only does index provide us with an example of an essentially normal operator which is not of the form normal plus compact, index is the only other ingredient needed to classify essentially normal operators. This is the main result for essentially normal operators.

THEOREM 2. *Two essentially normal operators* T_1 *and* T_2 *are unitarily equivalent* modulo $\mathcal{K}$ *if and only if* $\sigma_e(T_1) = \sigma_e(T_2) = X$ and $\operatorname{ind}(T_1-\lambda) = \operatorname{ind}(T_2-\lambda)$ *for* λ *in* $\mathbb{C}\setminus X$.

Now what does this have to do with extensions? If $\mathcal{E}_T$ denotes the C^*-subalgebra of $\mathcal{L}(\mathfrak{H})$ generated by I, T and $\mathcal{K}(\mathfrak{H})$, then the quotient $\mathcal{E}_T/\mathcal{K}(\mathfrak{H})$ is the C^*-subalgebra of $\mathcal{Q}(\mathfrak{H})$ generated by 1 and $\pi(T)$. Since $\mathcal{E}_T/\mathcal{K}(\mathfrak{H})$ is commutative we have

$$\mathcal{E}_T/\mathcal{K}(\mathfrak{H}) \cong C(\sigma_{\mathcal{Q}(\mathfrak{H})}(\pi(T))) = C(\sigma_e(T))$$

by the spectral theorem and we have an extension

$$0 \longrightarrow \mathcal{K}(\mathfrak{H}) \longrightarrow \mathcal{E}_T \xrightarrow{\phi_T} C(\sigma_e(T)) \longrightarrow 0\,,$$

where ϕ_T is the restriction of π to $\mathcal{E}_T$. Moreover, one can show that the two extensions $(\mathcal{E}_{T_1}, \phi_{T_1})$ and $(\mathcal{E}_{T_2}, \phi_{T_2})$ are equivalent (in a sense which we make clear in the next chapter) if and only if T_1 and T_2 are unitarily equivalent modulo $\mathcal{K}$. Furthermore, if $\mathcal{K}(\mathfrak{H}) \subset \mathcal{E} \subset \mathcal{L}(\mathfrak{H})$ and $X \subset \mathbb{C}$ are such that

$$0 \longrightarrow \mathcal{K}(\mathfrak{H}) \xrightarrow{i} \mathcal{E} \xrightarrow{\phi} C(X) \longrightarrow 0$$

is exact, then any T in $\mathcal{E}$ is essentially normal. Moreover, if $\phi(T) = z$, then $\sigma_e(T) = X$ and $(\mathcal{E}_T, \phi_T)$ is equivalent to $(\mathcal{E}, \phi)$. Thus all extensions of $\mathcal{K}$ by $C(X)$ arise from essentially normal operators for $X \subset \mathbb{C}$. Therefore the problem of classifying essentially normal operators is equivalent to that of classifying extensions.

Thus there are various reasons to be interested in extensions of $\mathcal{K}$ by $C(X)$. In our study of these extensions the equivalence classes for fixed X are shown to form an abelian group $\mathrm{Ext}(X)$ such that $X \mapsto \mathrm{Ext}(X)$ defines a homotopy invariant covariant functor. Moreover, this functor can be used to define a generalized Steenrod homology theory which is dual to K-theory. Let me conclude this chapter by showing how this proves the theorem stated earlier.

One of the pairings we define between Ext and K-theory yields a homomorphism

$$\gamma_\infty : \mathrm{Ext}(X) \to \mathrm{Hom}(K^1(X), \mathbf{Z})$$

which we show is an isomorphism for $X \subset \mathbf{C}$. In this special case we will write γ_1 for this homomorphism. Since $X \subset \mathbf{C}$ we have

$$K^1(X) = H^1(X, \mathbf{Z}) = \pi^1(X) ,$$

where $\pi^1(X)$ is the first cohomotopy group of X or the group relative to pointwise multiplication of homotopy classes of maps from X into the nonzero complex numbers $\mathbf{C}^*$. The definition of

$$\gamma_1 : \mathrm{Ext}(X) \to \mathrm{Hom}(\pi^1(X), \mathbf{Z})$$

goes as follows: fix $(\mathcal{E}, \phi)$ in $\mathrm{Ext}(X)$ and for $f: X \to \mathbf{C}^*$ we define

$$\gamma_1(\mathcal{E}, \phi)[f] = \mathrm{ind}(\phi^{-1}(f)) ,$$

where $[f]$ denotes the element of $\pi^1(X)$ defined by f. To see that this is well defined, note that $\phi^{-1}(f)$ is not a unique Fredholm operator but has a well-defined index which depends only on the homotopy class of f. It is easy to check that γ_1 is a homomorphism. Now for $X \subset \mathbf{C}$ what is $\pi^1(X)$? If we let $\mathbf{C} \setminus X = O_\infty \cup O_1 \cup O_2 \cup \cdots$ denote the components, where O_∞ is the unbounded one, then $\pi^1(X)$ is the free abelian group with one generator for each bounded component. For the extension $(\mathcal{E}_T, \phi_T)$, the homomorphism $\gamma_1(\mathcal{E}_T, \phi_T)$ is defined by $[O_i] \to n_i$, where $n_i = \mathrm{ind}(T - \lambda_i)$ for some λ_i in O_i, and now Theorem 2 is obvious. Moreover, since γ_1 is surjective, the equivalence classes of essentially normal operators with essential spectrum X are obtained by prescribing arbitrary integers for the bounded components of the complement of X in C.

Equivalently, since

$$\mathrm{Hom}(\pi^1(X), Z) \cong \mathrm{Hom}(H^1(X,Z), Z) \cong H^0(C \setminus X, Z) \cong [C \setminus X, Z]$$

by Steenrod duality (where $[C \setminus X, Z]$ denotes the group of locally constant integer-valued functions defined on $C \setminus X$) γ_1 can be defined by

$$\gamma_1(\mathcal{E}_T, \phi_T)(\lambda) = \mathrm{ind}(T-\lambda I)$$

and Theorem 2 is obvious.

Although our proof certainly involves operator theory, it also involves a critical use of ideas and techniques from algebraic topology and homological algebra. Moreover, there is presently no proof of these results which does not.[1] In fact there is no purely operator theoretic proof of either of the following corollaries

COROLLARY. *An essentially normal operator* T *is in* $\mathfrak{N} + \mathcal{K}$ *if and only if* $\mathrm{ind}(T-\lambda) = 0$ *for* λ *in* $C \setminus \sigma_e(T)$.

COROLLARY. *The collection* $\mathfrak{N} + \mathcal{K}$ *is norm-closed.*

The situation in several variables is more subtle. For example, the analogue of the last corollary is false for compact perturbations of commuting pairs of normal operators. More precisely, the collection

$$\{N_1+K_1, N_2+K_2 : N_1, N_2 \in \mathfrak{N}, [N_1, N_2] = 0, K_1, K_2 \in \mathcal{K}\}$$

---

[1] In [30] Davie gives an exposition of the proofs of these results in which the ideas from operator theory and algebraic topology are separated as much as possible and the latter kept to a minimum.

is not norm closed. Why is that the case? Certain types of topological pathologies cannot occur in $\mathbf{C}$ but do in $\mathbf{C}^2$ (and all $\mathbf{C}^n$ for $n \geq 2$). Here we are using the fact that the generalized homology theory defined by Ext is not continuous. Thus Ext is sensitive enough to detect these pathologies.

The existence of the above pathology for pairs causes me to doubt that a purely operator theoretic proof of the last corollary is possible.

CHAPTER 2
Ext AS A GROUP

In this chapter we state the definitions with some care and describe how the basic results are proved. We shall indicate some generalizations and applications along the way.

For X a compact metrizable space and $\mathfrak{H}$ a separable complex Hilbert space we define an *extension* of $\mathcal{K}(\mathfrak{H})$ by $C(X)$ to be a pair $(\mathfrak{E}, \phi)$, where $\mathfrak{E}$ is a C^*-subalgebra of $\mathfrak{L}(\mathfrak{H})$ containing $\mathcal{K}(\mathfrak{H})$ and I, and ϕ is a *-homomorphism from $\mathfrak{E}$ onto $C(X)$ with kernel $\mathcal{K}(\mathfrak{H})$. Two extensions $(\mathfrak{E}_1, \phi_1)$ and $(\mathfrak{E}_2, \phi_2)$ are said to be (*strongly*) *equivalent* if there exists a *-isomorphism $\psi : \mathfrak{E}_1 \to \mathfrak{E}_2$ such that $\phi_2 \circ \psi = \phi_1$ and $\mathrm{Ext}(X)$ denotes the collection of equivalence classes of extensions of $\mathcal{K}(\mathfrak{H})$ by $C(X)$.

We make a few observations before continuing. The space X is required to be metrizable for two reasons. We must often use the fact that X and $C(X)$ are separable for sequential induction arguments. More importantly, unmanageable pathology can occur if X is allowed to be non-metrizable (cf. [20]). Perhaps one should investigate extensions for such X by allowing the ideal of compact operators on non-separable Hilbert spaces or even other ideals. Secondly, there is a difference between our definition of extension and the definition which obtains by allowing $\mathfrak{E}$ to be an abstract C^*-algebra and requiring $i: \mathcal{K} \to \mathfrak{E}$ to be a *-monomorphism. The difference is not serious and is explained in our earlier paper [20]. Moreover, unless a weaker notion of equivalence were used, Ext would not be a group if we used this broader definition. Since the *-isomorphism ψ in the definition of equivalence restricts to a *-automorphism on $\mathcal{K}(\mathfrak{H})$ and since all *-automorphisms of $\mathcal{K}(\mathfrak{H})$ are weakly inner, there exists a unitary U on $\mathfrak{H}$ such that ψ is the restriction of the inner automorphism a_U on $\mathfrak{L}(\mathfrak{H})$ defined by $a_U(X) = U^*XU$. We add that in homological algebra one would require that ψ restrict to the identity on $\mathcal{K}(\mathfrak{H})$. If we did that "Ext" would then be an extension of our Ext by the infinite unitary group. Our results would seem to justify our definition, at least in this context.

There is an equivalent definition of extension which is the "working definition". If we consider the commutative diagram

$$\begin{array}{ccccccccc} 0 & \longrightarrow & \mathcal{K}(\mathfrak{H}) & \longrightarrow & \mathcal{E} & \xrightarrow{\phi} & C(X) & \longrightarrow & 0 \\ & & \| & & \downarrow & & \downarrow{\scriptstyle \tau} & & \\ 0 & \longrightarrow & \mathcal{K}(\mathfrak{H}) & \longrightarrow & \mathfrak{L}(\mathfrak{H}) & \xrightarrow{\pi} & \mathfrak{Q}(\mathfrak{H}) & \longrightarrow & 0 \end{array}$$

then we obtain the *-monomorphism τ defined by $\pi \circ \phi^{-1}$. Conversely, if τ is a *-monomorphism of C(X) into $\mathfrak{Q}(\mathfrak{H})$, then the above diagram can be completed, where $\mathcal{E} = \pi^{-1}[\mathrm{im}\,\tau]$ and $\phi = \tau^{-1} \circ \pi$. Thus Ext(X) can also be defined as equivalence classes of *-monomorphisms of C(X) into $\mathfrak{Q}(\mathfrak{H})$, where $\tau_1 \sim \tau_2$ if $\tau_2 = a_{\pi(U)}\tau_1$ for some unitary U on $\mathfrak{H}$. In this context it would seem that equivalence would be more naturally defined by a_u, where u is a unitary element of $\mathfrak{Q}(\mathfrak{H})$ or a_u is an inner *-automorphism on $\mathfrak{Q}(\mathfrak{H})$. For extensions of $\mathcal{K}(\mathfrak{H})$ by C(X), both notions of equivalence are the same. Thus for $\mathcal{C}$ a separable commutative C*-algebra, we can define $\mathrm{Ext}(\mathcal{C})$ to be the equivalence classes of *-monomorphisms of $\mathcal{C}$ into $\mathfrak{Q}(\mathfrak{H})$ with equivalence being defined by inner automorphisms. It is natural to extend this definition to consider extensions of $\mathcal{K}(\mathfrak{H})$ by $\mathcal{C}$, where $\mathcal{C}$ is a general separable C*-algebra. However, for general $\mathcal{C}$ the two notions of equivalence do not agree. For example $\mathrm{Ext}(M_n) = 0$, while $\mathrm{Ext}^{str}(M_n) = Z_n$, where M_n denotes the C*-algebra of complex $n \times n$ matrices. While "Ext" defined for both notions of equivalence seems useful, we shall concentrate almost exclusively on that defined by weak equivalence on the collection of unital

*-monomorphisms of $\mathcal{C}$ into $\mathfrak{Q}(\mathfrak{H})$ which we shall denote by Ext. Before continuing we should comment on a question which the use of inner automorphisms raises; namely what about outer automorphisms on $\mathfrak{Q}(\mathfrak{H})$? It is an unsolved problem whether they exist and even if they do, it seems quite possible that their action on a separable subalgebra of $\mathfrak{Q}(\mathfrak{H})$ could be duplicated by an inner automorphism.

Now guided by homological algebra we define addition in Ext and the notion of trivial extension. In homological algebra it is straightforward to show that the equivalence classes form a commutative semigroup and that the trivial extension is unique up to equivalence and acts as an additive identity. In our context the properties of trivial extensions lie considerably deeper. We shall develop certain notions for general $\mathcal{C}$ but not all; indeed the theory is not complete for general C^*-algebras. The generalization of Ext to a larger class of C^*-algebras was dictated by my work with Brown and Fillmore and the general outline for arbitrary C^*-algebras was set forth by Brown [19]. Some of the necessary notions were based on [55].

Let τ_1 and τ_2 be *-monomorphisms from $\mathcal{C}$ into $\mathfrak{Q}(\mathfrak{H})$ and $a_1 = [\tau_1]$ and $a_2 = [\tau_2]$ denote the elements of $\mathrm{Ext}(\mathcal{C})$ they determine. Further, let $\rho : \mathfrak{Q}(\mathfrak{H}) \oplus \mathfrak{Q}(\mathfrak{H}) \to \mathfrak{Q}(\mathfrak{H})$ be the map determined by the diagram

$$\begin{array}{ccccc} \mathfrak{L}(\mathfrak{H}) \oplus \mathfrak{L}(\mathfrak{H}) & \longrightarrow & \mathfrak{L}(\mathfrak{H} \oplus \mathfrak{H}) & \xrightarrow{\nu} & \mathfrak{L}(\mathfrak{H}) \\ \big\downarrow{\scriptstyle \pi \oplus \pi} & & & & \big\downarrow{\scriptstyle \pi} \\ \mathfrak{Q}(\mathfrak{H}) \oplus \mathfrak{Q}(\mathfrak{H}) & & \xrightarrow{\quad\rho\quad} & & \mathfrak{Q}(\mathfrak{H}) \end{array}$$

where ν is induced by any unitary between $\mathfrak{H} \oplus \mathfrak{H}$ and $\mathfrak{H}$. Now if $\tau : \mathcal{C} \to \mathfrak{Q}(\mathfrak{H})$ is the map defined by

$$\tau(x) = \rho(\tau_1(x) \oplus \tau_2(x))$$

for x in $\mathcal{C}$, then we set $a_1 + a_2 = [\tau]$. One must, of course, show that this does not depend on ν or the choice of representatives; it does not and $\mathrm{Ext}(\mathcal{C})$ becomes a commutative semigroup.[2] It may be of interest to note that

$$[(\mathcal{E}_{T_1}, \phi_{T_1})] + [(\mathcal{E}_{T_2}, \phi_{T_2})] = [(\mathcal{E}_{T_1 \oplus T_2}, \phi_{T_1 \oplus T_2})]$$

for extensions defined by essentially normal operators.

An extension $\tau : \mathcal{C} \to \mathcal{Q}(\mathfrak{H})$ is said to be *trivial* if there exists a unital $*$-monomorphism $\sigma : \mathcal{C} \to \mathcal{L}(\mathfrak{H})$ such that $\tau = \pi \circ \sigma$ or equivalently, if in the following diagram τ lifts:

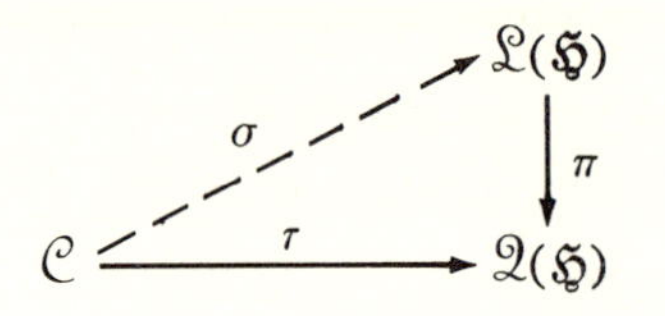

The pair $(\mathcal{E}, \phi)$ defines a trivial extension if there exists a $*$-monomorphism $\sigma : C(X) \to \mathcal{E}$ such that $\phi \circ \sigma = \mathrm{id}_{C(X)}$.[3] If $(\mathcal{E}_T, \phi_T)$ is trivial, then there exists $\sigma : C(\sigma_e(T)) \to \mathcal{E}_T$; hence $\sigma(z) = N$ is a normal operator and $\phi(T) = z = \phi(N)$ which implies $T - N$ is compact. Thus $(\mathcal{E}_T, \phi_T)$ trivial implies T is a normal operator plus a compact. Conversely, if we can write $T = N + K$, then a little work using the spectral theorem shows that it is possible to choose N such that $\sigma(N) = \sigma_e(T)$ and the spectral theorem provides $\sigma : C(\sigma(N)) \to \mathcal{E}_T$. Thus $(\mathcal{E}_T, \phi_T)$ is trivial if and only if T is a normal plus compact.

---

[2] The topologist will recognize this as the analogue of Whitney sum for vector bundles: if $X \to BU(m)$ and $X \to BU(n)$ define vector bundles over X, then this sum is defined by the composite $X \to BU(m) \times BU(n) \to B(U(m) \times U(n)) \to B(U(m+n))$, where $BU(n)$ denotes the classifying space for the unitary group $U(n)$.

[3] The topologist will again note the similarity: a bundle over X defined by $X \to BU(n)$ is trivial precisely when the map lifts to the (contractible) total space $EU(n)$ of the universal $BU(n)$-bundle

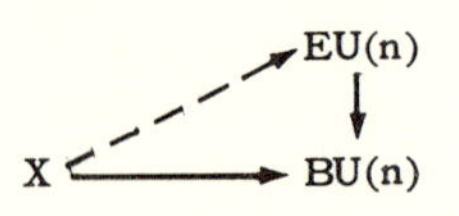

THEOREM 3. *For* T *essentially normal the extension* $(\mathcal{E}_T, \phi_T)$ *is trivial if and only if* $T = N + K$, *where* N *is normal and* K *is compact.*

For $X \subset \mathbb{C}$ if $\tau_1, \tau_2 : C(X) \to \mathfrak{Q}(\mathfrak{H})$ define trivial extensions, then there exist normal operators N_1 and N_2 such that $\sigma_e(N_i) = X$ and $\pi(N_i) = \tau_i(z)$ for $i = 1, 2$. Therefore, by the Weyl-von Neumann-Berg theorem there exists a unitary operator U such that $N_1 - U^* N_2 U$ is compact; hence $\tau_1 = \alpha_{\pi(U)} \tau_2$ or $[\tau_1] = [\tau_2]$. Thus the Weyl-von Neumann-Berg theorem implies that there exists a unique trivial element in $\operatorname{Ext}(X)$ for $X \subset \mathbb{C}$. This situation for general X is contained in the following:

THEOREM 4. *For each compact metrizable space* X *there exists a unique trivial element in* $\operatorname{Ext}(X)$.

Proof. To exhibit a trivial element let $\{x_n\}$ be a countable dense subset of X, where each point occurs infinitely often. Let $\sigma : C(X) \to \mathfrak{L}(\ell^2)$, where $\sigma(f)$ is defined to be the diagonal operator on ℓ^2 with entries $\{f(x_n)\}$. Then $\tau = \pi \circ \sigma$ defines a trivial element of $\operatorname{Ext}(X)$. Moreover, two such extensions based on the sequences $\{x_n\}$ and $\{y_n\}$, respectively, are equivalent by the unitary operator defined by a permutation ρ on $\mathbb{N}$ satisfying $d(x_n, y_{\rho(n)}) \to 0$. It is easy to show that such a permutation exists. Thus uniqueness rests on showing that every trivial extension is equivalent to such an extension.

To do that let $\tau : C(X) \to \mathfrak{Q}(\mathfrak{H})$ be a trivial extension and $\sigma : C(X) \to \mathfrak{L}(\mathfrak{H})$ be defined such that $\tau = \pi \circ \sigma$. By the spectral theorem there exists a spectral measure $E(\cdot)$ defined on the Borel subsets of X such that

$$\sigma(f) = \int_X f dE \quad \text{for } f \text{ in } C(X).$$

If $\{U_n\}$ is a basis for the open subsets of X and $\mathfrak{B}$ denotes the uniformly closed subalgebra of the bounded Borel functions on X generated by

$\{X_{U_n}\}$, then $C(X) \subset \mathfrak{B} \cong C(\tilde{X})$ by the Gelfand-Naimark theorem. Moreover, $\tilde{X}$ is totally disconnected since $C(\tilde{X})$ is generated by idempotents. Also there exists a surjective map $\eta: \tilde{X} \to X$ dual to the inclusion $C(X) \subset C(\tilde{X})$, and σ can be extended using the spectral measure $E(\cdot)$ to obtain $\tilde{\sigma}: C(\tilde{X}) \to \mathfrak{L}(\mathfrak{H})$. If ψ is a homeomorphism of $\tilde{X}$ onto a subset Λ of R, then $H = \tilde{\sigma}(\psi)$ is self-adjoint and hence by the Weyl-von Neumann theorem there exists a diagonal operator $D = \operatorname{diag}\{\lambda_n\}$ with $\sigma(D) = \Lambda$ such that $D-H$ is compact. Using the spectral theorem a *-monomorphism $\nu: C(\Lambda) \to \mathfrak{L}(\mathfrak{H})$ can be defined such that

$$\nu(f) = \operatorname{diag}\{f(\lambda_n)\}.$$

If we set $x_n = \eta[\psi^{-1}(\lambda_n)]$ and define $\sigma_o(g) = \operatorname{diag}\{g(x_n)\}$ for g in $C(X)$, then $[\tau] = [\pi \circ \sigma_o]$ which completes the proof.

The proof that the trivial element in $\operatorname{Ext}(X)$ acts as the identity on $\operatorname{Ext}(X)$ requires some additional ideas. A result due to Fillmore, Stampfli and Williams [38] shows that if λ is in $\sigma_e(T)$ for T on $\mathfrak{H}$, then we can decompose $\mathfrak{H} = \mathfrak{H}_1 \oplus \mathfrak{H}_2$ such that T is unitarily equivalent modulo $\mathcal{K}(\mathfrak{H})$ to the operator defined by the matrix

$$\begin{pmatrix} \lambda I & 0 \\ 0 & T' \end{pmatrix}$$

where T' is an operator on $\mathfrak{H}_2$. By iterating this result one can show that T is unitarily equivalent modulo $\mathcal{K}(\mathfrak{H})$ to the operator defined on $\mathfrak{H} = \mathfrak{H}_3 \oplus \mathfrak{H}_4$ by the matrix

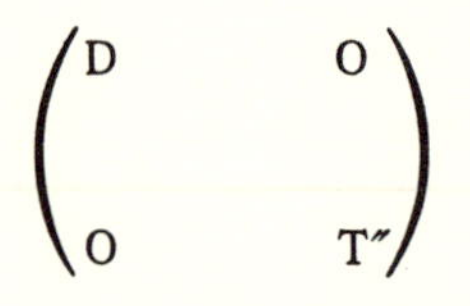

where D is a diagonal operator with $\sigma_e(D) = \sigma_e(T)$ and T'' is some operator.

If $\tau: C(X) \to \mathfrak{Q}(\mathfrak{H})$ is a *-monomorphism for $X \subset \mathbf{C}$, then applying the preceding to an operator T such that $\pi(T) = \tau(z)$ the *-monomorphism $\sigma: C(X) \to \mathfrak{L}(\mathfrak{H}_3)$ such that $\sigma(z) = D$ and the *-monomorphism $\tau': C(X) \to \mathfrak{Q}(\mathfrak{H}_4)$ such that $\tau'(z) = \pi(T'')$ are obtained. Then if $\tau_X : C(X) \to \mathfrak{Q}(\mathfrak{H}')$ defines a trivial extension, we have

$$[\tau] + [\tau_X] = [\tau'] + [\pi \circ \sigma] + [\tau_X] = [\tau'] + [\pi \circ \sigma] = [\tau] ,$$

using the fact that the sum of two trivial extensions is trivial.

For general X we embed X as a subset of $\prod_{n \geq 1} [0,1]$ and let $\{\psi_n\}$ be the coordinate maps. If $\tau: C(X) \to \mathfrak{Q}(\mathfrak{H})$ defines an extension, then the previous technique is applied to a tuple $\{T_n\}$ of operators chosen such that $\tau(\psi_n) = \pi(T_n)$. Thus we are able to prove that:

THEOREM 5. *The unique trivial element of* $\operatorname{Ext}(X)$ *acts as an identity and thus* $\operatorname{Ext}(X)$ *is a commutative semigroup with identity.*

Before continuing we want to show how these results plus a little operator theory enable us to calculate $\operatorname{Ext}(\mathbf{T})$ for $\mathbf{T}$ the unit circle in $\mathbf{C}$. If $(\mathcal{E}_T, \phi_T)$ defines an element in $\operatorname{Ext}(\mathbf{T})$, then using the polar decomposition and the spectral theorem one can show that $T = V + K$, where V is an isometry, a co-isometry, or a unitary depending upon whether $\operatorname{ind}(T) = -n$ is positive, negative or zero respectively. In the first case the von Neumann-Wold decomposition asserts that V is unitarily equivalent to $V_n \oplus W$, where $V_n = T_{z^n}$ and W is a unitary. By appealing to the last theorem we see that a further compact perturbation yields that $(\mathcal{E}_T, \phi_T)$ is equivalent to $(\mathcal{E}_{V_n}, \phi_{V_n})$. A similar argument for the case $n < 0$ and the Weyl-von Neumann theorem for the case $n = 0$ yields that $(\mathcal{E}_T, \phi_T)$ is determined by $\operatorname{ind} T$. Therefore, we have proved that $\operatorname{Ext}(\mathbf{T}) = \mathbf{Z}$.

Theorem 5 has been extended by Voiculescu [99] to general separable C^*-algebras. An improved exposition of his proof was given by Arveson

[5]. These results have some interesting consequences which we will describe in part. For noncommutative $\mathcal{C}$ the maximal ideal space must be replaced by the space of irreducible representations of $\mathcal{C}$. If $\{\nu_n\}$ denotes a family of irreducible representations of $\mathcal{C}$ such that $\oplus\nu_n$ is faithful, then we can obtain a trivial extension of $\mathcal{K}$ by $\mathcal{C}$ by considering $\pi\circ(\oplus(\nu_n\otimes I))$, where I is the identity operator on an infinite dimensional separable Hilbert space. Voiculescu's results state that up to weak equivalence, the unital trivial extension of $\mathcal{K}$ by $\mathcal{C}$ is unique and acts as the identity in $\operatorname{Ext}(\mathcal{C})$. This implies that if $\nu: \mathcal{C} \to \mathcal{L}(\mathfrak{H})$ is any representation of $\mathcal{C}$, then we can find a sequence of irreducible representations $\{\nu_n\}$ of $\mathcal{C}$ with $\nu_n: \mathcal{C} \to \mathcal{L}(\mathfrak{H}_n)$, where $\mathfrak{H} = \oplus\mathfrak{H}_n$, such that

$$\pi\circ(\nu(x)) = \pi\circ(\oplus\nu_n(x)) \text{ for } x \text{ in } \mathcal{C}.$$

Moreover, he also proves that one can make $\oplus\nu_n$ as close to ν in the point norm topology as desired. In particular, if T is an operator on $\mathfrak{H}$ and $\mathcal{C}_T$ is the C*-algebra generated by T, then the identity representation for $\mathcal{C}_T$ can be approximated in the strong topology by the direct sum of infinitely many irreducible representations. Hence, every operator T can be approximated in norm by reducible operators, answering in the affirmative the question of Halmos [41]. Voiculescu also proves that for $\mathcal{C}$ a separable C*-algebra of $\mathcal{Q}(\mathfrak{H})$, we have $\mathcal{C}'' = \mathcal{C}$, where $'$ denotes the relative commutant in $\mathcal{Q}(\mathfrak{H})$. A further consequence is that modulo the ideal of compact operators, all representations of separable C*-algebras split into the direct sum of irreducible representations. Hence neither direct integrals nor multiplicity theory is necessary if one is able to work modulo $\mathcal{K}$.

Returning now to the commutative case, another consequence of the existence of the identity is that for $\psi: X \to Y$ the map $\psi_*: \operatorname{Ext}(X) \to \operatorname{Ext}(Y)$ can be defined such that

COROLLARY. *The correspondence* $X \mapsto \operatorname{Ext}(X)$ *is a covariant functor.*

If $\psi: X \to Y$ is surjective, then the map $\psi^*: C(Y) \to C(X)$ defined by $\psi^*(f) = f(\psi)$ is injective and hence we can define $\psi_*: \mathrm{Ext}(X) \to \mathrm{Ext}(Y)$ by $\psi_*([\tau]) = [\tau(\psi^*)]$ for the *-monomorphism $\tau: C(X) \to \mathfrak{Q}(\mathfrak{H})$. If ψ^* is not surjective, then $\tau(\psi^*)$ is not a monomorphism. However, we can define

$$\psi_*([\tau]) = [\tau(\psi^*) \oplus \tau_Y],$$

where $\tau_Y: C(Y) \to \mathfrak{Q}(\mathfrak{K})$ defines the trivial extension. The properties necessary to be a covariant functor follow easily.

Now the most surprising thing about Ext (when we were first studying it and still today) is the fact that Ext(X) is an abelian group. In no situation were algebra extensions known to form a group and we were led to expect it only because Ext(X) was a group for the cases we could calculate. Our original proof of this was quite complicated and came after the construction of most of the other machinery in [20]. Fortunately, an easier proof was discovered by Arveson [4]. In fact, this latter proof was subsequently extended by Choi and Effros [27] to the class of separable nuclear C^*-algebras.

If $\tau: C(X) \to \mathfrak{Q}(\mathfrak{H})$ defines an extension, then $[\tau]$ is trivial if and only if τ lifts to a *-monomorphism $\sigma: C(X) \to \mathfrak{L}(\mathfrak{H})$ such that $\tau = \pi \circ \sigma$. In a more general context Andersen [1], and Vesterstøm [98] showed that there always exists a positive map ρ such that $\tau = \pi \circ \rho$. Arveson observed that this was enough for him to produce an inverse for $[\tau]$ in Ext(X). The argument goes as follows. A theorem due to Naimark asserts that for such a map ρ there exists a Hilbert space $\mathfrak{K}$ containing $\mathfrak{H}$ and a *-homomorphism $\psi: C(X) \to \mathfrak{L}(\mathfrak{K})$ such that $\rho(f) = P_{\mathfrak{H}}\psi(f)P_{\mathfrak{H}}$ for f in C(X). If we decompose

$$\psi = \begin{pmatrix} \psi_{11} & \psi_{12} \\ \psi_{21} & \psi_{22} \end{pmatrix} \text{ relative to } \mathfrak{K} = \mathfrak{H} \oplus \mathfrak{H}^\perp, \text{ then both } \pi \circ \psi_{11} = \tau$$

and $\pi \circ \psi$ are homomorphisms. An easy calculation shows that $\pi \circ \psi_{12} = 0$ and $\pi \circ \psi_{21} = 0$ and hence $\pi \circ \psi_{22}$ is a *-homomorphism from C(X) to

$\mathfrak{Q}(\mathfrak{H}^{\perp})$. If we define $[\tau'] = [\pi \circ \psi_{22} \oplus \tau_X]$, where $\tau_X : C(X) \to \mathfrak{L}(\mathfrak{G})$ defines the trivial extension, then $[\tau']$ is the inverse for $[\tau]$ in Ext(X).[4]

THEOREM 6. *The correspondence* $X \mapsto \mathrm{Ext}(X)$ *is a covariant functor from the category of compact metrizable spaces and continuous maps to the category of abelian groups and homomorphisms.*

Actually the class of maps $\rho : \mathcal{C} \to \mathfrak{L}(\mathfrak{H})$ for which Naimark's theorem is valid is the class of completely positive ones. The map ρ is said to be *completely positive* if $\rho \otimes 1 : \mathcal{C} \otimes M_n \to \mathfrak{L}(\mathfrak{H}) \otimes M_n$ is positive for each n. For a commutative C^*-algebra $\mathcal{C}$ positive maps are completely positive but this is false for general $\mathcal{C}$. A C^*-algebra $\mathcal{C}$ is *nuclear* if the identity map on $\mathcal{C}$ can be approximated in the strong topology by completely positive maps which factor through M_n for some n. Arveson reproved [5] (also cf. [100]) the Choi-Effros result [27] that $\mathrm{Ext}(\mathcal{C})$ is a group for $\mathcal{C}$ a separable nuclear C^*-algebra by showing first that the completely positive maps from $\mathcal{C}$ to $\mathfrak{Q}(\mathfrak{H})$ which lift to completely positive maps from $\mathcal{C}$ to $\mathfrak{L}(\mathfrak{H})$ are closed in the strong topology and second by invoking Choi's result (cf. [5]) that all completely positive maps lift for $\mathcal{C} = M_n$.

It would be nice if $\mathrm{Ext}(\mathcal{C})$ were a group for all separable C^*-algebras $\mathcal{C}$ but that is not the case. Anderson [3] showed that $\mathrm{Ext}(\mathcal{C})$ is not a group, where $\mathcal{C}$ is a C^*-algebra closely related to the group C^*-algebra for the free group F_2 on two generators. (However, Ext of the group C^*-algebra for F_2 is a group.) This could be troublesome for some possible applications to topology where the relevant C^*-algebra would seem to be the group C^*-algebra for the fundamental group of a manifold.

One last comment before concluding this chapter. From the existence of inverses it follows that all extensions can be obtained from a Toeplitz-

---

[4]The topologist will again note the analogy. Given a vector bundle ξ, to produce its inverse one embeds ξ in a trivial bundle $\mathcal{O}^n$. The sequence of vector bundles $0 \to \xi \to \mathcal{O}^n \to \mathcal{O}^n/\xi \to 0$ splits (since X is compact), so there is a subbundle ξ' of $\mathcal{O}^n$ with $\xi \oplus \xi' = \mathcal{O}^n$. Then ξ' is an inverse for ξ in $\tilde{K}^0(X)$, that is, modulo trivial bundles.

like construction. If $\tau: C(X) \to \mathfrak{Q}(\mathfrak{H})$ is a *-monomorphism and $\tau^{-1}: C(X) \to \mathfrak{Q}(\mathfrak{H}_0)$ defines an inverse, then

$$\tau \oplus \tau^{-1}: C(X) \to \mathfrak{Q}(\mathfrak{H} \oplus \mathfrak{H}_0)$$

is trivial and hence can be lifted to a map $\sigma: C(X) \to \mathfrak{L}(\mathfrak{H} \oplus \mathfrak{H}_0)$. Since all *-monomorphisms $\sigma: C(X) \to \mathfrak{L}(\mathfrak{H} \oplus \mathfrak{H}_0)$ are unitarily equivalent modulo $\mathfrak{K}(\mathfrak{H} \oplus \mathfrak{H}_0)$ we can assume that σ is unitarily equivalent to a multiplier representation on an L^2 space defined on X. Thus we can assume that there exists a measure μ on X such that $\mathfrak{H} \oplus \mathfrak{H}_0 \cong L^2(\mu)$ and $\sigma(\psi) = M_\psi$ for ψ in $C(X)$. If we do that then τ is defined by

$$\tau(\psi) = \pi \circ (P_{\mathfrak{H}} M_\psi | \mathfrak{H}) \quad \text{for } \psi \text{ in } C(X) .$$

Hence all elements of $\operatorname{Ext}(X)$ are obtained as in the Toeplitz construction by compressing a multiplier representation $\sigma(\psi) = M_\psi$ of $C(X)$ on an L^2 space to a subspace $\mathfrak{H}$. The condition which such a subspace $\mathfrak{H}$ must satisfy is that the projection $P_{\mathfrak{H}}$ must essentially commute with the M_ψ for ψ in $C(X)$. Such subspaces arise naturally from an operator F which essentially commutes with the representation, if the essential spectrum of F has two components. Each component determines a subspace of the L^2 space which essentially commutes. One way for this to happen is for F to be essentially self-adjoint and Fredholm; then the positive essential spectrum determines an element of $\operatorname{Ext}(X)$ while the negative determines its inverse.

CHAPTER 3
Ext AS A HOMOTOPY FUNCTOR

In the first two chapters we have defined $X \mapsto \operatorname{Ext}(X)$ and shown it to be a covariant functor from the category of compact metrizable spaces to the category of abelian groups. In this chapter we derive the first piece of the exact sequence for Ext and the fact that Ext is a homotopy functor. We define the negatively indexed Ext groups using suspension and show that Ext defines a generalized homology theory except for the existence of the positively indexed groups. The definition of these must wait for the periodicity theorem which will be obtained later.

Before discussing the exact sequence or homotopy invariance, we describe our basic technique for dealing with Ext. Let $\tau: C(X) \to \mathfrak{Q}(\mathfrak{H})$ be a *-monomorphism and suppose that somehow we have found a *commutative* C^*-algebra $\mathfrak{Z} \subset \mathfrak{Q}(\mathfrak{H})$ which contains $\operatorname{im}\tau$. If $\tilde{X}$ denotes the maximal ideal space of $\mathfrak{Z}$ and we define the *-monomorphism $\alpha: C(X) \to C(\tilde{X})$ by the diagram

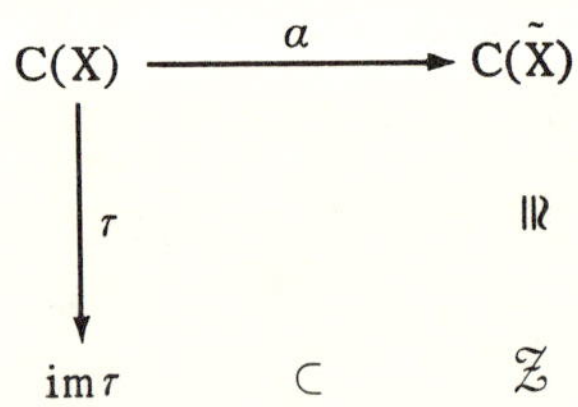

then there exists a surjective map $\psi: \tilde{X} \to X$ such that $\alpha(f) = f \circ \psi = \psi^*(f)$. Moreover, if we define $\tilde{\tau}: C(\tilde{X}) \to \mathfrak{Q}(\mathfrak{H})$ by the diagram

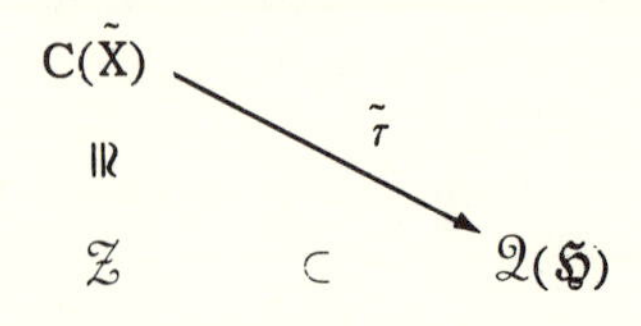

then τ defines an element of $\mathrm{Ext}(\tilde{X})$ such that $\psi_*[\tilde{\tau}] = [\tau]$. Thus if we know something about $[\tilde{\tau}]$ in $\mathrm{Ext}(\tilde{X})$, then we have information about $[\tau]$.[5] For example, if $\mathcal{Z}$ is generated by projections, then $\tilde{X}$ is totally disconnected and as we saw in the proof of the theorem in the last chapter, this implies $\mathrm{Ext}(\tilde{X}) = 0$ and hence $[\tau] = 0$. If $\mathcal{Z}$ is generated by $\mathrm{im}\,\tau$ and some projection p, then $\tilde{X}$ is the disjoint union $B \vee C$ of B and C, where B and C are closed subsets of X (where the Gelfand transform $\hat{p}$ of p is 0 and 1, respectively) which satisfy $B \cup C = X$ and $\psi: B \vee C \to X$ is just projection. Since $\mathrm{Ext}(B \vee C) = \mathrm{Ext}(B) \oplus \mathrm{Ext}(C)$, this enables us to replace questions about $\mathrm{Ext}(X)$ by questions concerning $\mathrm{Ext}(B)$ and $\mathrm{Ext}(C)$. The precise details are contained in the Mayer-Vietoris sequence toward which these arguments lead. The difficulty in this approach lies in finding an appropriate $\mathcal{Z}$. One method is contained in the following

LEMMA. *Let* $p: X \to Y$ *be surjective, let* $\tau: C(X) \to \mathfrak{Q}(\mathfrak{H})$ *define an extension such that* $p_*(\tau) = 0$, *let* $\sigma: C(X) \to \mathfrak{L}(\mathfrak{H})$ *be such that* $p_*(\tau) = \tau(p^*) = \pi \circ \sigma$, *and let* $E(\cdot)$ *be the spectral measure on* Y *satisfying*

$$\sigma(f) = \int_Y f\,dE \quad \text{for } f \text{ in } C(Y).$$

If C *is a closed subset of* Y *such that* ∂C *contains no point with multiple preimage, then* $\pi(E(C))$ *commutes with* $\mathrm{im}\,\tau$. *Moreover, the maximal ideal space* $\underline{X}$ *of the algebra* $\mathcal{Z}$ *generated by* $\mathrm{im}\,\tau$ *and* $\pi(E(C))$ *is* $p^{-1}(C) \vee p^{-1}(Y \setminus C)$.

The proof is straightforward once one observes that the functions f in $C(X)$ of the form $f = f_1 \circ p + f_2 + f_3$, where f_1 is in $C(Y)$, f_2 is

[5]The topologist will note the analogy (with interesting contrasts) with the splitting principle for vector bundles.

supported on $p^{-1}(\text{int}(C))$ and f_3 is supported on $p^{-1}(Y \setminus C)$, are dense in $C(X)$. Since each point in C has a unique preimage, $f \circ p^{-1}$ defines a continuous function on ∂C which can be extended to a function f_1 on Y by the Tietze extension theorem. The function $f - f_1 \circ p$ vanishes on ∂C and hence can be uniformly approximated by a function vanishing in a neighborhood of ∂C. Thus we obtain f_2 and f_3. Since $\tau(f_1 \circ p) = \pi(\sigma(f_1))$ we see that $\pi(E(C))$ commutes with $\tau(f_1 \circ p)$. For f_2 choose g in $C(Y)$ supported on C such that $(g \circ p) \cdot f_2 = f_2 \cdot (g \circ p)$. Then we obtain

$$\begin{aligned}\pi(E(C))\tau(f_2) &= \pi(E(C))\tau(g \circ p)\tau(f_2) = \tau(g \circ p)\tau(f_2) = \tau(f_2)\tau(g \circ p) \\ &= \tau(f_2)\tau(g \circ p)\pi(E(C)) = \tau(f_2)\pi(E(C)) .\end{aligned}$$

The argument for f_3 is analogous.

An example for which this lemma is useful is the theta curve. Let X be the theta curve lying on its side, Y be the right half, and p be reflection in the vertical line of symmetry.

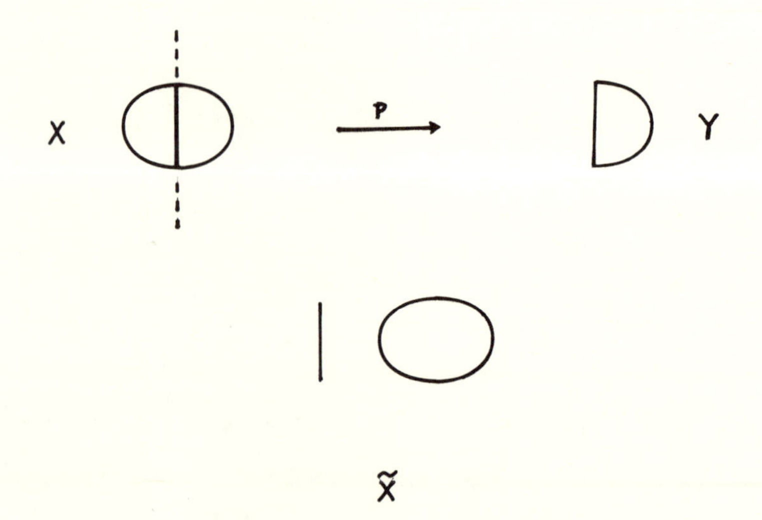

If we let C denote the vertical piece, then ∂C consists of the two end points on the preimage of which p is one-to-one. Thus $\tau : C(X) \to \mathfrak{Q}(\mathfrak{H})$ can be promoted to $\tilde{\tau}: C(\tilde{X}) \to \mathfrak{Q}(\mathfrak{H})$ if and only if the sum of the indices corresponding to the two regions is equal to 0. The index to which we are referring is $\gamma_1(\tau)$ evaluated at the element of $\pi^1(X)$ corresponding to the region, where the map $\gamma_1 : \mathrm{Ext}(X) \to \mathrm{Hom}(\pi^1(X), \mathbf{Z})$ is that defined in the first chapter. Since $\mathrm{Ext}(\tilde{X}) = \mathbf{Z}$, we obtain $\mathrm{Ext}(X) = \mathbf{Z} \oplus \mathbf{Z}$.

Adding one projection is usually not enough. In the following lemma we add enough projections to totally disconnect the rest of the space.

LEMMA. *Let* $p: X \to Y$ *be surjective and let* B *be a closed subset of* Y *containing all points with multiple preimage in* X. *Let* $A = p^{-1}(B)$, *let* $p': A \to B$ *be the restriction of* p, *and let* $i: A \to X$ *and* $j: B \to Y$ *be inclusion maps. Then* $\ker p_* \subset i_*(\ker p'_*)$.

If we consider the diagram

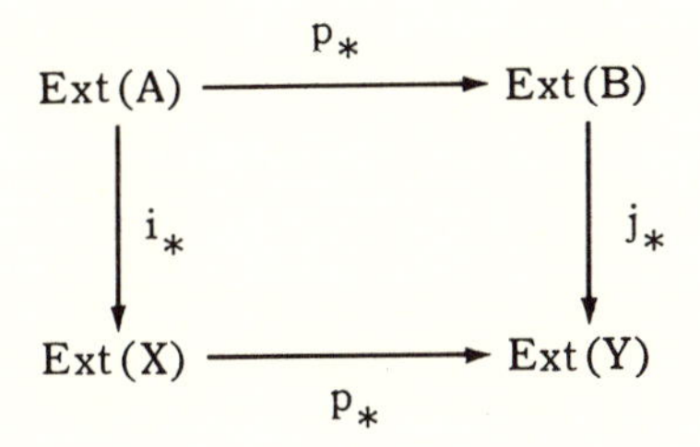

then the lemma says that if τ in $\mathrm{Ext}(X)$ maps to 0 in $\mathrm{Ext}(Y)$, then τ is the image of an element in $\mathrm{Ext}(A)$ which maps to 0 in $\mathrm{Ext}(B)$.

Proof. Let $\tau: C(X) \to \mathfrak{Q}(\mathfrak{H})$ be such that $p_*(\tau) = 0$, and let σ and $E(\cdot)$ be as in the previous lemma. Select a basis $\{U_k\}$ for the topology of $X \setminus A$ so that $C_k = \bar{U}_k$ is disjoint from A and let $\mathcal{Z}$ be the commutative C^*-subalgebra of $\mathfrak{Q}(\mathfrak{H})$ generated by $\mathrm{im}\,\tau$ and the projections $\pi(E(p(C_k)))$. Note that $p(C_k)$ is a subset of Y such that the hypotheses of the preceding lemma are satisfied. If $\tilde{X}$ denotes the maximal ideal space of $\mathcal{Z}$ and $u: \tilde{X} \to X$, then u is a homeomorphism on $u^{-1}(A)$ which we identify with A with inclusion map $\tilde{i}: A \to \tilde{X}$ and $\tilde{X}/A$ is totally disconnected. Moreover, we have $\tilde{\tau}: C(\tilde{X}) \to \mathcal{Q}(\mathfrak{H})$ such that $u_*(\tilde{\tau}) = \tau$.

One can prove that if we have $A \to \tilde{X} \to \tilde{X}/A$, where the latter space $\tilde{X}/A$ is totally disconnected, then $\tilde{i}_*: \mathrm{Ext}(A) \to \mathrm{Ext}(\tilde{X})$ is surjective. Thus we have $\tau': C(A) \to \mathcal{Q}(\mathfrak{H})$ such that $\tilde{i}_*(\tau') = \tilde{\tau}$. But $u \circ \tilde{i} = i$ and hence $i_*(\tau') = \tau$. To show that $p'_*[\tau'] = 0$, we observe that $\mathrm{im}(\tau' \circ p'^*)$ is contained in the range of $\pi(E(\cdot))$ and as in the preceding chapter, if the image of an extension is contained in an algebra generated by projections, then it is trivial.

Before continuing let us remark that the fact that the map $\tilde{i}_*$ is surjective when $\tilde{X}/A$ is totally disconnected is not trivial unless $\tilde{X}/A$ is the disjoint union of finitely many open and closed subsets. Otherwise it involves a kind of infinite sum (cf. [20, 6.6]), which is possible since $\tilde{X}/A$ can be written as the infinite union of open and closed sets with diameter converging to zero.

THEOREM 7. *If* A *is a closed subset of the compact metrizable space* X, *then*

$$\mathrm{Ext}(A) \xrightarrow{i_*} \mathrm{Ext}(X) \xrightarrow{p_*} \mathrm{Ext}(X/A)$$

is exact, where $i: A \to X$ *is inclusion and* $p: X \to X/A$ *is the quotient map.*

We have $\ker p_* \subset \mathrm{im}\, i_*$ by the previous lemma and the other inclusion is obvious since $\mathrm{Ext}(\text{point}) = 0$.

If $X = B \cup C$ and $A = B \cap C$, then considering the diagram

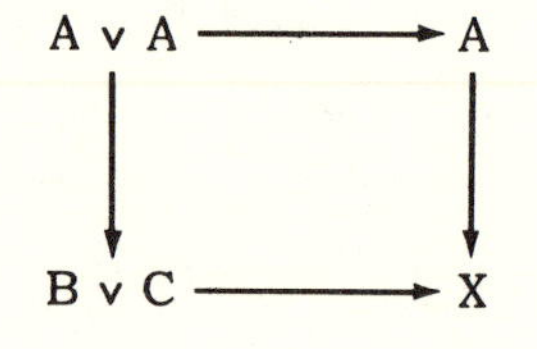

and applying the previous lemma gives the first part of the Mayer-Vietoris sequence for Ext:

$$\mathrm{Ext}(A) \to \mathrm{Ext}(B) \oplus \mathrm{Ext}(C) \to \mathrm{Ext}(X)\ .$$

In a recent paper [79] Voiculescu and two of his former students Pimsner and Popa have extended Theorem 7 to the class of separable nuclear C^*-algebras. This has very important consequences when combined with homotopy invariance allowing all the results of [22] including periodicity to be extended to the class of type I separable C^*-algebras.

By virtue of Theorem 7, to show Ext is homotopy invariant it suffices to prove that $\mathrm{Ext}\, CX = 0$ for each X, where CX denotes the *cone* $CX = X \times I/X \times \{0\}$ over X with $I = [0,1]$. If $f, g\colon X \to Y$ are homotopic and $i, j\colon X \to X \times I$ are the inclusions on $X \times \{0\}$ and $X \times \{1\}$, respectively, then there exists $F\colon X \times I \to Y$ such that $f = F \circ i$ and $g = F \circ j$. Now

$$\mathrm{Ext}(X) \to \mathrm{Ext}(X \times I) \to \mathrm{Ext}(CX)$$

is exact by the theorem, and if $\mathrm{Ext}(CX) = 0$, then i_* is surjective. Thus given τ in $\mathrm{Ext}(X \times I)$, there exists ν in $\mathrm{Ext}(X)$ such that

$$i_*(\nu) = j_*(\tau)\ .$$

But the projection $p\colon X \times I \to X$ satisfies $p \circ i = p \circ j = \mathrm{id}_X$, and hence $\nu = \tau$. Therefore $i_* = j_*$ which implies

$$f_* = F_* \circ i_* = F_* \circ j_* = g_*\ .$$

Thus Ext is homotopy invariant if we can show that Ext of the cone over each space is 0.

This part of the proof makes use several times of the technique of showing $\mathrm{Ext}(X) = 0$ by replacing X by a subdivided X, iterating, taking an inverse limit and hence replacing X by a totally disconnected space.

Recall that the *inverse limit* of a sequence of spaces $\{X_n\}$ and maps $f_n: X_{n+1} \to X_n$ $(n \geq 1)$ is defined to be

$$\lim_{\leftarrow} X_n = \left\{ x \in \prod_{n=1}^{\infty} X_n : f_n(x_{n+1}) = x_n, n \geq 1 \right\}$$

with the subspace topology. Analogously, for a sequence of groups $\{G_n\}$ and homomorphisms $\rho_n: G_{n+1} \to G_n$ $(n \geq 1)$ the inverse limit is defined to be

$$\lim_{\leftarrow} G_n = \left\{ g \in \prod_{n=1}^{\infty} G_n : \rho_n(g_{n+1}) = g_n, n \geq 1 \right\}.$$

One always has the homomorphism

$$P: \operatorname{Ext}(\lim_{\leftarrow} X_n) \to \lim_{\leftarrow} \operatorname{Ext}(X_n)$$

defined by $P(\tau) = \{p_{k*}(\tau)\}$, where $p_k: \lim_{\leftarrow} X_n \to X_k$ is the coordinate projection.

THEOREM 8. *If* $\{(X_n, f_n)\}$ *is an inverse limit of compact metrizible spaces and continuous maps, then the induced map* P *is surjective.*

This homomorphism is not always injective which means that the homology theory defined by Ext is not continuous [34]. The kernel of P can be characterized (cf. Chapter 5). Also we will only indicate the proof assuming that the f_n are all surjective since that is the generality we need. (The general case is only slightly harder.)

Proof. Let $\{a_n\}$ define an element of $\lim_{\leftarrow} \operatorname{Ext}(X_n)$, that is, a_n is in $\operatorname{Ext}(X_n)$ and $f_{n*}(a_{n+1}) = a_n$ for all $n \geq 1$. Choose $\tau_n: C(X_n) \to \mathfrak{Q}(\mathfrak{H})$ such that τ_n defines a_n and $\tau_{n+1} \circ f_n^* = \tau_n$ as follows:

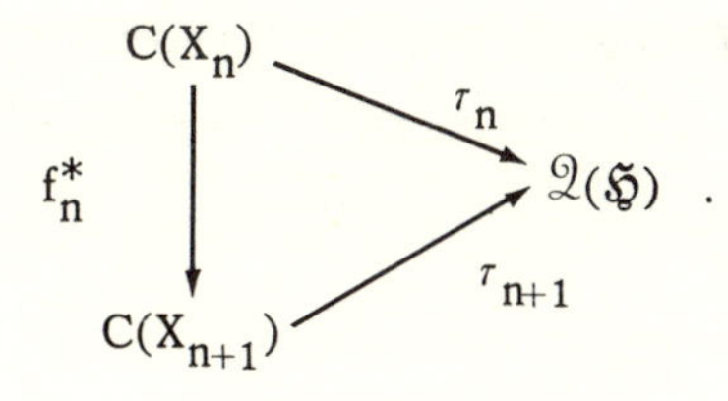

Suppose we have chosen a $\tau_1 : C(X_1) \to \mathfrak{Q}(\mathfrak{H})$ and $\tau_2 : C(X_2) \to \mathfrak{Q}(\mathfrak{H})$ such that $a_1 = [\tau_1]$ and $a_2 = [\tau_2]$. Then τ_1 and $\tau_2 | C(X_1)$ are equivalent and hence there exists a unitary element u of $\mathfrak{Q}(\mathfrak{H})$ such that

$$\alpha_u \tau_2 | C(X_1) = \tau_1 .$$

Then take $\alpha_u \tau_2$ for τ_2 and continue. Collectively, the $\{\tau_n\}$ define a $*$-monomorphism on the dense subalgebra

$$\cup\, p_n^*[C(X_n)]$$

of $C(\underleftarrow{\lim}\, X_n)$ which can be extended to $\tau : C(\underleftarrow{\lim}\, X_n) \to \mathfrak{Q}(\mathfrak{H})$. But $P(\tau) = \{a_n\}$ which completes the proof.

LEMMA. *For any sequence* $\{X_n\}$ *of closed subsets of* $[0, 1]$,

$$\mathrm{Ext}\left(\prod_{n=1}^{\infty} X_n\right) = 0 .$$

Proof. Let $X = \prod_{n \geq 1} X_n$, $X' = \prod_{n \geq 2} X_n$, $B = (X_1 \cap [0, \frac{1}{2}]) \times X'$ and $C = (X_1 \cap [\frac{1}{2}, 1]) \times X'$. Since $B \cap C$ is a retract of C, one can show that

$$\mathrm{Ext}(B \vee C) \to \mathrm{Ext}(X)$$

is surjective. Now iterating this we obtain a sequence of spaces $\{Y_n\}$ and maps $p_n : Y_{n+1} \to Y_n$ such that $Y_1 = X, Y_2 = B \vee C$, p_{n*} is surjective, and each Y_n consists of a finite number of closed subsets of X the maximum diameter of which goes to 0. Thus $\lim_{\leftarrow} Y_n$ is totally disconnected, $\mathrm{Ext}(\lim_{\leftarrow} Y_n) = 0$, and hence by the theorem we have $\lim_{\leftarrow} \mathrm{Ext}(Y_n) = 0$. Since each p_{n*} is surjective this implies $\mathrm{Ext}(Y_n) = 0$ for $n \geq 1$ or $\mathrm{Ext}(X) = 0$.

COROLLARY. $\mathrm{Ext}(I^k) = 0$, $\mathrm{Ext}(I^\omega) = 0$ *and* $\mathrm{Ext}(CI^\omega) = 0$.

The last statement requires observing that CI^ω imbeds as a convex subset of Hilbert space and hence is an absolute retract. The proof then follows by imbedding CI^ω in the Hilbert cube I^ω.

The fact that $\mathrm{Ext}(I^2) = 0$ yields implications in operator theory which have not been proved by other means. For example if $L: \ell^2 \to \ell^2$ is defined by

$$Le_n = \lambda_n e_{n+1}$$

where the λ_n are $1, \frac{1}{2}, 1, \frac{3}{4}, \frac{1}{2}, \frac{1}{4}, \frac{1}{2}, \frac{3}{4}, 1, \frac{7}{8}, \cdots$, then one shows easily that $[L, L^*]$ is compact and $\sigma_e(L) = \bar{D}$. Since $\bar{D}$ and I^2 are homeomorphic, it follows that $\mathrm{Ext}(\bar{D}) = 0$ and therefore L lies in $\mathfrak{N} + \mathcal{K}$. No direct proof of this fact is known to me.

There is a class of essentially normal Toeplitz operators which one can show is actually in $\mathfrak{N} + \mathcal{K}$. If $\psi: T \to C$ is chosen such that the winding number of every point relative to the curve determined by ψ is 0, then T_ψ lies in $\mathfrak{N} + \mathcal{K}$ but again no direct proof is known.

We complete now our unfinished business. The argument that $\mathrm{Ext}(CX) = 0$ proceeds as follows. Let $X = X_1 \cup X_2$ and let $p: CX_1 \vee CX_2 \to CX$ be the obvious "projection". Using ideas suggested by A. M. Davie (cf. [22], p. 279-280) involving the construction of various ingenious topological maps, one can show that p_* is surjective. Iterate this step always decomposing the subsets of X. Form an inverse system, the inverse limit of which this time is $I \times \tilde{X}$, where $\tilde{X}$ is totally disconnected. Then there is a surjection of $\tilde{X}$ onto X. By the previous lemma, $\mathrm{Ext}(I \times \tilde{X}) = 0$ and as in the proof of that lemma we have $\mathrm{Ext}(CX) = 0$. Combining this with our earlier remarks we have

THEOREM 9. *The correspondence* $X \mapsto \operatorname{Ext}(X)$ *is a homotopy functor; that is, if* $f, g: X \to Y$ *are homotopic, then* $f_*, g_*: \operatorname{Ext}(X) \to \operatorname{Ext}(Y)$ *are equal.*

COROLLARY. *If* X *is contractible, then* $\operatorname{Ext}(X) = 0$.

The fact that Ext is homotopy invariant has been extended to a much larger class of C^*-algebras. The combined work of O'Donovan [65] and Salinas [84] and more recently Pimsner, Popa and Voiculescu [79] show that $\operatorname{Ext}(\mathcal{C})$ is homotopy invariant for the class of separable type I C^*-algebras. A C^*-algebra $\mathcal{C}$ is said to be CCR if the range of every irreducible representation of $\mathcal{C}$ in $\mathfrak{L}(\mathfrak{H})$ is contained in $\mathcal{K}(\mathfrak{H})$. Further, it is said to be GCR if it has a composition series of ideals $\{\mathfrak{J}_\alpha\}$ indexed by ordinals such that $\mathfrak{J}_\alpha/\mathfrak{J}_{\alpha+1}$ is CCR for each α. For separable $\mathcal{C}$, type I and GCR coincide.

There is another way of looking at the subdivision technique for showing that an extension is trivial. If the map $\tau: C(X) \to \mathfrak{Q}(\mathfrak{H})$ defines a trivial extension, then there exists a spectral measure $E(\cdot)$ defined on the Borel subsets of X such that $\tau(f) = \pi\{\int_X f dE\}$ for f in $C(X)$. Conversely, if we can define such a spectral measure, then τ is seen to be trivial. Actually, we need considerably less than a measure defined on all Borel subsets since to integrate continuous functions the Riemann integral suffices. If $X = B \cup C$ and there exist $\tau_1 : C(B) \to \mathfrak{Q}(\mathfrak{H}_1)$ and $\tau_2 : C(C) \to \mathfrak{Q}(\mathfrak{H}_2)$ such that $i_{B*}(\tau_1) + i_{C*}(\tau_2) = \tau$, then there exists a projection p in $\mathfrak{Q}(\mathfrak{H})$ commuting with $\operatorname{im} \tau$ such that the maximal ideal space of the algebra generated by $\operatorname{im} \tau$ and p is $B \vee C$. Pull p back to a projection P in $\mathfrak{L}(\mathfrak{H})$. Assign P to B and $I-P$ to $X \backslash B$ and proceed. So long as the diameter of the pieces into which we subdivide X goes to 0, then we obtain a projection-valued measure on sufficiently many subsets of X to define the Riemann integral and integration yields $\sigma: C(X) \to \mathfrak{L}(\mathfrak{H})$ such that $\tau = \pi \circ \sigma$. Therefore τ is trivial.

There is another technique for showing homotopy invariance due to O'Donovan [65] and Salinas [84] which forms the basis for the generalizations to larger classes of C^*-algebras. As above, homotopy invariance reduces to showing that every extension τ in $\mathrm{Ext}(CX)$ is trivial. This is accomplished by constructing an extension τ' such that $[\tau]+[\tau'] = [\tau']$ and hence from the group structure of $\mathrm{Ext}(CX)$ it follows that $[\tau] = 0$. The construction takes two steps. First the C^*-algebra $\mathfrak{E} = \pi^{-1}[\mathrm{im}\,\tau]$ is shown to be quasi-diagonal and secondly, one exhibits τ'. A C^*-algebra $\mathfrak{E} \subset \mathfrak{L}(\mathfrak{H})$ is said to be *quasi-diagonal* if there exists an orthogonal sequence $\{P_n\}$ of finite dimensional projections such that $\Sigma P_n = I$ and $T - \Sigma P_n T P_n$ is compact for T in $\mathfrak{E}$. This notion was introduced for a single operator by Halmos [41] and its relevance to the study of extensions was realized by O'Donovan [65]. Its use was one of the central ideas in Arveson's exposition [5] of the Voiculescu and Choi-Effros papers [99], [27].

To show that $\mathfrak{E}$ is quasi-diagonal one takes an inverse τ'' of τ and completely positive liftings ρ and ρ'' of τ and τ'' to $\mathfrak{L}(\mathfrak{H})$ and $\mathfrak{L}(\mathfrak{H}'')$, respectively. Using the maps $\Theta_\lambda : CX \to CX$ defined by $\Theta_\lambda(t, x) = (\lambda t, x)$ for $0 \le \lambda \le 1$, a sequence of *-monomorphisms $\sigma_n : C(CX) \to \mathfrak{L}(\mathfrak{H} \oplus \mathfrak{H}'')$ is produced such that

$$\|\sigma_n(f) - \rho(f)\| \to 0 \quad \text{for } f \text{ in } C(CX) .$$

From this follows the existence of a sequence of orthogonal projections $\{P_k\}$ relative to which $\mathrm{im}\,\rho$ is quasi-diagonal. An inductive argument along the lines of the proof of the Weyl-von Neumann theorem (cf. [20]) can be used. The definition of τ' is given in terms of the decomposition $\{P_n\}$ using direct sums of $P_n(\Theta_{\lambda^*} \circ \rho) P_n$. Rather than give this argument which is notationally quite complicated we shall specialize to the case where $CX = \overline{D}$, adapting O'Donovan's argument [65] slightly.

Since $CX = \overline{D}$ is a subset of C, we must show that every essentially normal operator T with $\sigma_e(T) = \overline{D}$ lies in $\mathfrak{N} + \mathfrak{K}$. The first step in this argument follows Pearcy and Salinas [70] who showed that such an operator is quasi-triangular.

Let S be an essentially normal operator such that $\sigma_e(S) = \overline{D}$ and $T \oplus S = N + K$. Then for each n we have

$$T \oplus \left(\bigoplus_{k=0}^{n-1} \frac{k}{n} N \right) = T \oplus \left(\bigoplus_{k=0}^{n-1} \frac{k}{n} T \right) \oplus \left(\bigoplus_{k=0}^{n-1} \frac{k}{n} S \right) + K'$$

$$= \left(\bigoplus_{k=1}^{n} \frac{k}{n} T \right) \oplus \left(\bigoplus_{k=1}^{n} \frac{k}{n} S \right) \oplus (0) -$$

$$0 \oplus \left(\bigoplus_{k=1}^{n} \frac{1}{n} S \right) \oplus 0 + K'$$

$$= \left(\bigoplus_{k=0}^{n} \frac{k}{n} N \right) - X + K'' ,$$

where K', K'' are compact and $\|X\| = \|\bigoplus_{k=1}^{n} \frac{1}{n} S\| = \frac{1}{n} \|S\|$. Since $T \oplus \left(\bigoplus_{k=0}^{n-1} \frac{k}{n} N \right)$ is unitarily equivalent to a compact perturbation of T, it follows that T is in the closure of $\mathfrak{N} + \mathfrak{K}$. (This would imply T itself lies in $\mathfrak{N} + \mathfrak{K}$ by the result stated in Chapter 1 except our argument would be circular. Also in the case of general X, what we obtain is that the lifted map p is the point limit of *-monomorphisms.) Using Halmos' result [41], we see that T is quasi-diagonal or $T = \oplus T_i + K$, where T_i is defined on H_i and $\dim H_i < \infty$. To complete the proof set $R = \bigoplus_{j=1}^{\infty} \bigoplus_{i=0}^{2^j-1} \frac{i}{2^j} T_j \oplus 0$, where the zero term is infinite dimensional. Since $[T, T^*] = \oplus [T_i, T_i^*] + K'$ is compact, it follows that $\|[T_i, T_i^*]\| \to 0$ and hence $[R, R^*] = \bigoplus_{j=1}^{\infty} \bigoplus_{i=0}^{2^j-1} \frac{i}{2^{j+1}} [T_j, T_j^*]$ is compact. Therefore R is essentially normal and since we can assume $\|R\| \leq 1$, we have $\sigma_e(R) \subset \overline{D}$. Finally, since

$$T \oplus R = \bigoplus_{j=1}^{\infty} \bigoplus_{i=0}^{2^j-1} \frac{i+1}{2^j} T_j \oplus 0 ,$$

where all the zero terms have been amalgamated, there is a unitary operator W such that $T \oplus R - W^*RW = \bigoplus_{j=1}^{\infty} \bigoplus_{i=0}^{2^j-1} \frac{1}{2^j} T_j \oplus 0$ which is compact. This completes the proof.

CHAPTER 4

GENERALIZED HOMOLOGY THEORY AND PERIODICITY

So far we have shown that Ext defines a covariant homotopy functor such that for every closed subset $A \subset X$, we have the exact sequence

$$\operatorname{Ext}(A) \xrightarrow{i_*} \operatorname{Ext}(X) \xrightarrow{p_*} \operatorname{Ext}(X/A),$$

where $i: A \to X$ is inclusion and $p: X \to X/A$ is the quotient map. We define the higher reduced Ext groups by

$$\operatorname{Ext}_n(X) \equiv \operatorname{Ext}(S^{1-n}X) \quad \text{for} \quad n \leq 1,$$

where SX denotes the suspension $SX = CX/X \times \{1\}$ and S^k denotes iterated suspension. Then standard techniques from algebraic topology (e.g. [45]), using the Barratt-Puppe argument, allow one to prove

THEOREM 10. *For any closed subset* A *of* X, *there is a natural exact sequence*

$$\operatorname{Ext}_1(A) \longrightarrow \operatorname{Ext}_1(X) \longrightarrow \operatorname{Ext}_1(X/A) \xrightarrow{\partial} \operatorname{Ext}_0(A) \longrightarrow \operatorname{Ext}_0(X) \longrightarrow \cdots .$$

The maps $\operatorname{Ext}_n(A) \to \operatorname{Ext}_n(X)$ are defined by inclusion $A \subset X$ while the maps $\operatorname{Ext}_n(X) \to \operatorname{Ext}_n(X/A)$ are defined by the quotient map $X \to X/A$. We define the boundary maps $\partial: \operatorname{Ext}_n(X/A) \to \operatorname{Ext}_{n-1}(A)$ by $\partial = p_* q_*^{-1}$, where

$$p: X \cup CA \to (X \cup CA)/X = SA$$

and

$$q: X \cup CA \to (X \cup CA)/CA = X/A .$$

It is convenient to define the non-reduced groups for Ext, which we denote by $E_k(X)$, such that

$$E_1(X) = \mathrm{Ext}(X) \text{ and } E_k(X) = \ker \pi_* \text{ for } k \leq 0 ,$$

where π is the projection $\pi: X \times S^{1-k} \to X$.

A *generalized homology theory* is a homology theory satisfying all the Eilenberg-Steenrod axioms [34] except possibly for the dimension axiom. Standard techniques from algebraic topology show that except for the absence of the groups indexed by $k \geq 2$, we have a generalized homology theory. We define the missing groups by establishing the periodicity of Ext; that is, we define a natural map

$$\mathrm{Per}_*: \mathrm{Ext}(S^2 X) \to \mathrm{Ext}(X)$$

and show that it is an isomorphism for all X. This map corresponds to the Bott isomorphism in K-theory which we denote by $\mathrm{Per}^*: K^0(X) \to K^{-2}(X)$. However our proof of periodicity for Ext does not use the result from K-theory.

Before we can define Per_* we must establish some of the pairings between Ext and K-theory. We mentioned the pairing

$$\mathrm{Ext}_1(X) \to \mathrm{Hom}(K^1(X), Z)$$

or equivalently,

$$\mathrm{Ext}_1(X) \times K^1(X) \to Z$$

in the first chapter. We now consider it in more detail. Let $\tau: C(X) \to \mathfrak{Q}(\mathfrak{H})$ be a *-monomorphism. For each integer n consider the diagram

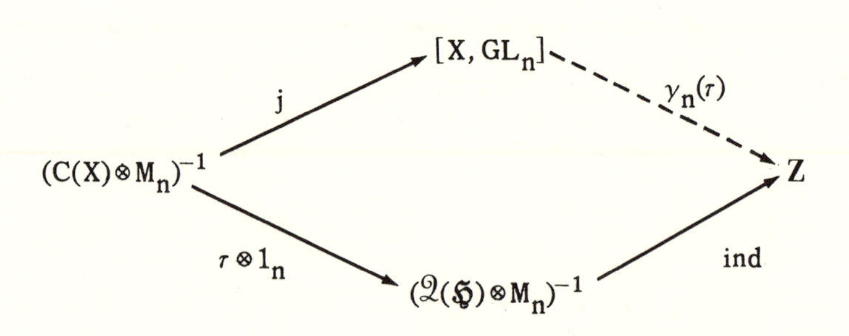

which defines the homomorphism $\gamma_n(\tau): [X, GL_n] \to Z$ as follows, where $[X, GL_n]$ denotes the group of homotopy classes of maps from X to $GL_n(C)$. Since we can identify $C(X) \otimes M_n$ as the algebra $M_n(X)$ of continuous M_n-valued functions on X, $(C(X) \otimes M_n)^{-1}$ is just the group of continuous GL_n-valued functions on X and the homomorphism j is defined by taking a function to its homotopy class. The homomorphism $\tau \otimes 1_n: C(X) \otimes M_n \to \mathfrak{Q}(\mathfrak{H}) \otimes M_n$ is obvious and identifying $\mathfrak{Q}(\mathfrak{H}) \otimes M_n$ with $\mathfrak{Q}(\mathfrak{H} \otimes C^n)$ yields the index homomorphism

$$\text{ind}: (\mathfrak{Q}(\mathfrak{H}) \otimes M_n)^{-1} \to Z .$$

Also simple properties of index show that $\gamma_n(\tau)$ is well defined. Moreover, if $\lambda: GL_n \to GL_{n+1}$ is the map defined by

$$\lambda(A) = \begin{pmatrix} A & 0 \\ 0 & 1 \end{pmatrix},$$

then the following diagram commutes:

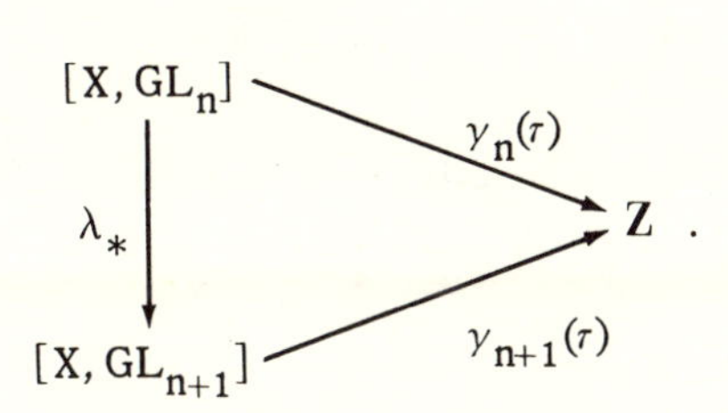

Since the direct limit of the groups $[X, GL_n]$ yields

$$K^1(X) = \lim_{\leftarrow} [X, GL_n] \quad \text{(cf. [6], [50], [96])} ,^6$$

we obtain the homomorphism $\gamma_\infty(\tau): K^1(X) \to Z$ and

[6] Actually $\lim_{\leftarrow} [X, GL_n]$ is more usually called $K^{-1}(X)$, but $K^1(X) \cong K^{-1}(X)$ by Bott periodicity.

$$\gamma_\infty : \operatorname{Ext}(X) \to \operatorname{Hom}(K^1(X), Z)$$

or

$$\operatorname{Ext}(X) \times K^1(X) \to Z$$

which is the first pairing.

As we said in the first chapter, γ_∞ is an isomorphism for $X \subset R^2$ which is the basic result used in the classification of essentially normal operators. Actually γ_∞ is an isomorphism for $X \subset R^3$ but not for $X \subset R^4$. For this we need to know that there is a torsion element a of order two in $\operatorname{Ext}(RP^2)$, where RP^2 is the real projective plane, and that RP^2 embeds in R^4. Then $a \neq 0$ but $\gamma_\infty(a) = 0$ since the range of γ_∞ is torsion free. (Also because $K^1(RP^2) = 0$.)

It may be of interest to note that a representative for a can be written down explicitly. If M is the Möbius band, C is the circle running down the center of M, $p : M \to M/\partial M$ is the quotient map, $\sigma : C(M) \to \mathfrak{L}(\mathfrak{H})$ is a $*$-monomorphism, and ϕ is a homeomorphism from S^1 onto C, then τ can be realized by defining

$$\tau(f) = \pi(\sigma(f \circ p) \oplus T_{(f \circ p \circ \phi)}) \text{ for } f \text{ in } C(M/\partial M),$$

where T_g is the Toeplitz operator with symbol g. The proof follows by considering the sequence

$$\begin{array}{ccccc} \operatorname{Ext}(\partial M) & \longrightarrow & \operatorname{Ext}(M) & \longrightarrow & \operatorname{Ext}(M/\partial M) \\ \| & & \| & & \\ Z & \longrightarrow & Z & & \end{array}$$

where $\operatorname{Ext}(M)$ can be identified with $\operatorname{Ext}(C) = Z$ since C is a retract of M, and the map from $\partial M \to C$ is $z \to z^2$. Therefore $Z \to Z$ is the map $n \to 2n$ and the extension we have defined is just the image in $\operatorname{Ext}(M/\partial M)$ of a generator in $\operatorname{Ext}(M)$. Since $M/\partial M$ is homeomorphic to RP^2 we have our torsion element of $\operatorname{Ext}(RP^2)$. If we embed RP^2 in C^2 and obtain operators T_1 and T_2 such that

$$\tau(z_i) = \pi(T_i) \text{ for } i = 1, 2 ,$$

then since all index data vanishes we can choose T_1 and T_2 to be normal. Moreover, we have $[T_1, T_2]$ compact. However, while it is impossible to find compact operators K_1 and K_2 such that

$$T_1 + K_1 \text{ and } T_2 + K_2 \text{ are normal and } [T_1 + K_1, T_2 + K_2] = 0 ,$$

we can so perturb the pair $T_1 \oplus T_1$ and $T_2 \oplus T_2$. It is difficult to imagine a purely operator theoretic proof of these facts or of any example exhibiting this behavior.

Additional information on γ_∞ has been obtained by L. G. Brown. In particular, he shows [18] that there is an exact sequence

$$0 \longrightarrow \operatorname{ext}(K^0(X), Z) \longrightarrow \operatorname{Ext}(X) \xrightarrow{\gamma_\infty} \operatorname{Hom}(K^1(X), Z) \longrightarrow 0 ,$$

which splits unnaturally, where "ext" denotes the usual group extensions of $K^0(X)$ by Z. Thus γ_∞ is always surjective. Further $\ker \gamma_\infty$ is the torsion subgroup when X is a finite complex, while to identify $\ker \gamma_\infty$ in general one must first topologize $\operatorname{Ext}(X)$. Brown shows that $\ker \gamma_\infty$ is the maximal compact subgroup of $\operatorname{Ext}(X)$. The topology is defined using the point norm topology on the collection of all *-monomorphisms from $C(X)$ into $\mathfrak{Q}(\mathfrak{H})$. The quotient of this space by conjugacy yields the topological group $\operatorname{Ext}(X)$, where the topology is not necessarily Hausdorff since the maps defining trivial extensions are not closed in this topology for sufficiently pathological X.

This is the case, for example, for the suspension of a solenoid (cf. [30]). The n-adic solenoid Σ_n can be identified as the inverse limit $\{S^1, \phi_k\}$ of a sequence of circles with maps $\phi_k(z) = z^n$ for all k. Using various techniques [48], it can be shown that $\operatorname{Ext}_r(\Sigma_n) = 0$ for r odd while $\operatorname{Ext}_r(\Sigma_n) = \hat{Z}_n/Z$ for r even, where $\hat{Z}_n$ is the group of n-adic integers, and the closure of the identity in $\operatorname{Ext}_r(\Sigma_n)$ is the whole group. Since an n-adic solenoid can also be realized as the intersection of a sequence of nested tori in $\mathbf{R}^3$, where each successive torus winds n times in its predecessor, it follows that its suspension $S\Sigma_n$ is homeomorphic to a subset of $\mathbf{R}^4$ and the trivial extensions are not closed.

The fact that the set of *-homomorphisms defining trivial extensions is not always closed is what is responsible for the failure of the analogue of the norm-closure of $\mathfrak{N}+\mathcal{K}$ in several variables. That is, if one considers the collection of pairs of operators (T_1, T_2) which can be written in the form $T_i = N_i + K_i$, where N_1 and N_2 are commuting normal operators and K_1 and K_2 are compact, then this collection is not closed in the operator norm. It is this which makes it difficult to construct even a heuristic operator theoretic proof that $\mathfrak{N}+\mathcal{K}$ is closed.

The second and more important pairing between Ext and K-theory makes $\mathrm{Ext}(X)$ into a $K^0(X)$-module. Our definition is a bit roundabout and requires some preparation.

Fix X and let E be an n-dimensional Hermitian vector bundle over X. This means each fiber E_x is isomorphic to $\mathbf{C}^n$ and is equipped with an inner product which varies continuously as a function on X. A map $\Phi: E \to E$ is a *bundle map* if $\Phi E_x \subset E_x$ and $\Phi|E_x$ is linear for x in X. Thus each Φ in the set $C_E(X)$ of bundle maps on E defines at each x in X an operator Φ_x on $\mathbf{C}^n$. Hence the norm $\|\Phi_x\|$ is defined and is continuous on X. Then, $C_E(X)$ is a C^*-algebra relative to the supremum norm defined by $\|\Phi\| = \sup\{\|\Phi_x\| : x \in X\}$ and involution defined pointwise. For $E = X \times \mathbf{C}^n$, $C_E(X)$ is just $M_n(X)$. We consider $\mathrm{Ext}(C_E(X))$ and hence *-monomorphisms

$$\tau' : C_E(X) \to \mathfrak{Q}(\mathfrak{H}) \text{ relative to } \textit{weak} \text{ equivalence.}$$

If $\tau: C(X) \to \mathfrak{Q}(\mathfrak{H})$ is a *-monomorphism, then tensoring by M_n yields

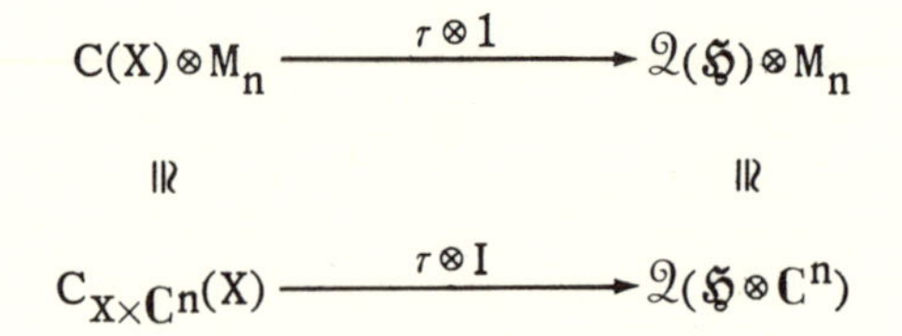

and hence a homomorphism $\mathrm{Ext}(X) \to \mathrm{Ext}(C_{X\times \mathbf{C}^n}(X))$ is defined which can be shown without much difficulty to be an isomorphism ([22], p. 292). This is closely related to the proof that $\mathrm{Ext}(M_n) = 0$ relative to weak equivalence and involves pulling back matrix units from $\mathfrak{Q}(\mathfrak{H})$ to $\mathfrak{L}(\mathfrak{H})$. This can always be done with a subspace of dimension less than n left over–hence the need for weak equivalence.

In [66] Olsen and Zame prove that if $X \subset \mathbf{C}^n$, then $C(X)\otimes M_n$ has a single generator. Therefore, if $(\mathcal{E}, \phi)$ is an extension of $\mathcal{K}$ by $C(X)$, then $\mathcal{E}\otimes M_n$ is singly generated modulo $\mathcal{K}$ and since $(\mathcal{E}\otimes M_n, \phi\otimes 1)$ determines $(\mathcal{E}, \phi)$, it follows that every extension can be "described" by a single operator. This enables one to use "Ext-phenomena" for higher dimensional spaces to obtain phenomena for n-normal operators. An *n-normal operator* is an $n\times n$ matrix with commuting normal entries (cf. Chapter 6). For example the two-torsion element in $\mathrm{Ext}(\mathbf{R}P^2)$ yields an essentially binormal operator which is not a compact perturbation of a binormal operator [33] defined on a subspace of finite codimension.

Further, if E_1 and E_2 are bundles over X such that E_1 is a subbundle of E_2, then there exists a natural homomorphism

$$i_* : \mathrm{Ext}(C_{E_2}(X)) \to \mathrm{Ext}(C_{E_1}(X))$$

which one defines as follows. There exists an orthogonal subbundle E_3 of E_2 such that $E_2 = E_1 \oplus E_3$ and a projection P_1 in $C_{E_1}(X)$ such that E_1 is the range of P_1. If τ is in $\mathrm{Ext}(C_{E_1}(X))$, then $p_1 = \tau(P_1)$ is a projection in $\mathfrak{Q}$. For Φ in $C_{E_1}(X)$ define $i^*(\Phi)$ in $C_{E_2}(X)$ to be Φ on E_1, 0 on E_3, and extend linearly on fibers. If we define $i_*\circ\tau$ such that $(i_*\circ\tau)(\Phi) = p_1\tau(i^*\circ\Phi)p_1$ then we obtain a *-monomorphism from $C_{E_1}(X)$ to $p_1\mathfrak{Q}p_1$. If P_1 is a projection in $\mathfrak{L}(\mathfrak{H})$ such that $\pi(P_1) = p_1$, then $p_1\mathfrak{Q}p_1$ can be identified with $\mathfrak{Q}(P_1\mathfrak{H})$ and thus $i_*\circ\tau$ defines an element of $\mathrm{Ext}(C_{E_1}(X))$.

Now suppose E_1 and E_2 are isomorphic subbundles of E such that $E_1^\perp$ and $E_2^\perp$ are isomorphic. If λ denotes the isomorphism between E_1 and E_2 and λ' the isomorphism between $E_1^\perp$ and $E_2^\perp$, then defining α to be λ on E, and λ' on $E^\perp$, and extending linearly to E yields an automorphism on E. Such an α is given by a unitary element in $C_E(X)$, and therefore $\alpha_* \circ \tau$ and τ are equivalent. Thus the following diagram commutes:

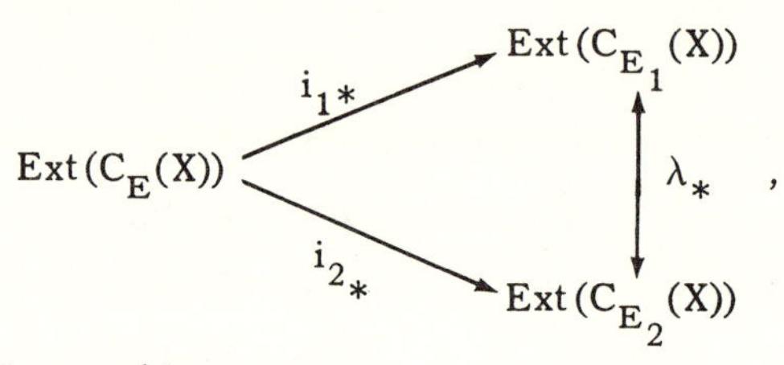

where λ_* is an isomorphism.

If we do not assume that $E_1^\perp$ and $E_2^\perp$ are isomorphic, then we replace E by $E \oplus E$. Since $(E \oplus E) \ominus (0 \oplus E_1) \cong E \oplus E_1^\perp \cong (E_2^\perp \oplus E_2) \oplus E_1^\perp \cong (E_2^\perp \oplus E_1) \oplus E_1^\perp \cong E_2^\perp \oplus E \cong (E \oplus E) \ominus (E_2 \oplus 0)$ we reduce to the previous case again obtaining that the diagram commutes and λ_* is an isomorphism. Now if E is isomorphic to any subbundle E_1 of $E \otimes C^n$, then the map $\operatorname{Ext}(C_{E \otimes C^n}(X)) \to \operatorname{Ext}(C_{E_1}(X))$ is an isomorphism since

$$E \cong E \oplus 0 \oplus \cdots \oplus 0 \subset E \oplus \cdots \oplus E \cong E \otimes C^n$$

is isomorphic to E_1.

THEOREM 11. *If* $E \xrightarrow{i} F$ *is a monomorphism of bundles over* X, *then* $\operatorname{Ext}(C_F(X)) \xrightarrow{i_*} \operatorname{Ext}(C_E(X))$ *is an isomorphism.*

Proof. In view of the diagram

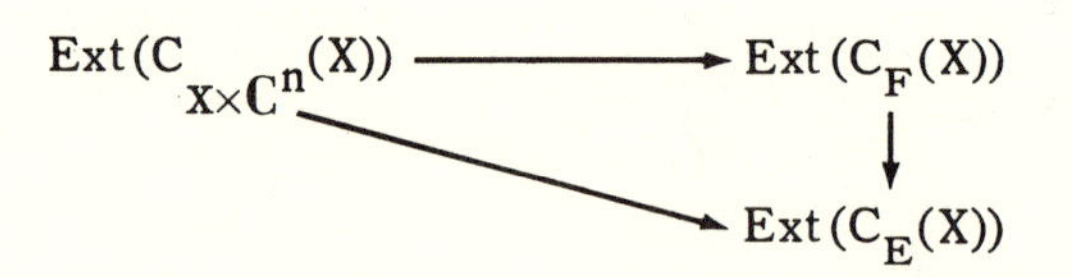

it is enough to consider the case when F is the trivial bundle over X, since we can imbed F in a trivial bundle. For any bundle E over X, we have $E \otimes \mathbf{C}^n$ contains the trivial bundle $X \times \mathbf{C}^k$ for n sufficiently large [45, Prop. 8.1.1]. And by our remarks prior to the statement of the theorem α and β are isomorphisms

$$\operatorname{Ext}(C_{X \times \mathbf{C}^{nk}}(X)) \to \operatorname{Ext}(C_{E \otimes \mathbf{C}^n}(X)) \to \operatorname{Ext}(C_{X \times \mathbf{C}^k}(X)) \to \operatorname{Ext}(C_E(X))$$

(α: from $\operatorname{Ext}(C_{E \otimes \mathbf{C}^n}(X))$ to $\operatorname{Ext}(C_E(X))$; β: from $\operatorname{Ext}(C_{X \times \mathbf{C}^{nk}}(X))$ to $\operatorname{Ext}(C_{X \times \mathbf{C}^k}(X))$)

from which the theorem follows.

As a consequence of this result we see that $\operatorname{Ext}(C_E(X))$ is an abelian group. A similar generalization to a larger class of C^*-algebras closely related to $C(X)$ was obtained by Salinas [83] before Choi and Effros obtained their result [27].

Now if E is a bundle over X, then it can be realized as a subbundle of a trivial bundle $X \times \mathbf{C}^n$ and by composing we obtain

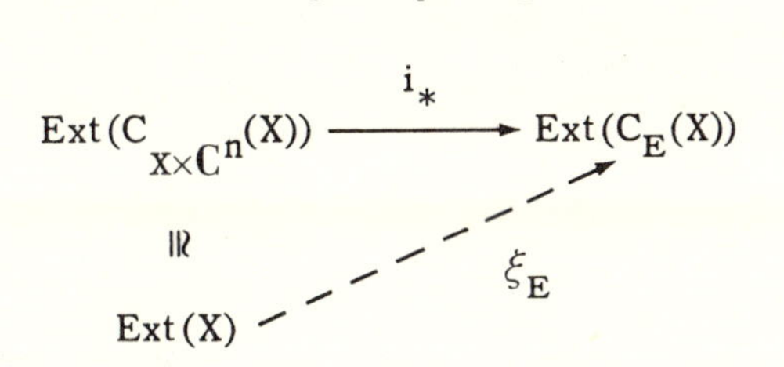

the natural isomorphism $\xi_E : \operatorname{Ext}(X) \to \operatorname{Ext}(C_E(X))$. Since the choice of $E \subset X \times \mathbf{C}^n$ is not natural, it is not immediately apparent that ξ_E is natural with respect to bundle maps $E \to E'$, but such is the case.

Now dual to the identification of $C(X)$ as the center of $C_E(X)$ there is a natural homomorphism

$$\zeta_E : \operatorname{Ext}(C_E(X)) \to \operatorname{Ext}(X)\,.$$

Finally, composition $m_E = \zeta_E \xi_E$ defines an endomorphism on $\operatorname{Ext}(X)$.

Thus we have the map

$$m : \mathrm{Vect}(X) \to \mathrm{End}(\mathrm{Ext}(X))$$

defined by $E \to m_E$. It can be shown that $m_{E\oplus F} = m_E + m_F$ and $m_{E\otimes F} = m_E \cdot m_F$. Since $K^0(X)$ is the Grothendieck ring completion of the semi-ring $\mathrm{Vect}(X)$ relative to $\oplus$ and $\otimes$, m extends to a ring homomorphism

$$m : K^0(X) \to \mathrm{End}(\mathrm{Ext}(X))$$

and hence $\mathrm{Ext}(X)$ becomes a $K^0(X)$-module. This pairing is natural in X. For example, since $\tilde{K}^0(RP^2) = Z_2$, then $\mathrm{Ext}(RP^2)$ is a Z_2-module.

Now although there are also two slant products between Ext and K-theory, we concentrate on the cap product since it is what is used to define Per_*. (We will discuss one of the slant products in the next chapter.) Before actually defining it we need other realizations of the higher E_*- and K^*-groups which one obtains by standard arguments. Let $S^{(0)}$ denote the one point space, $S^{(i)}$ be the product of i circles for $i > 0, p_j$ be the projection

$$p_j : X \times S^{(k)} \to X \times S^{(k-1)}$$

which omits the jth circle for $1 \leq j \leq k$, and q_j be the inclusion map

$$q_j : X \times S^{(k-1)} \to X \times S^{(k)}$$

which is obtained by choosing a base point on the jth circle.

LEMMA. *The group* $E_r(X)$ *is naturally isomorphic to the subgroup* $\bigcap_{j=1}^{1-r} \ker p_{j*}$ *of* $\mathrm{Ext}(X\times S^{(1-r)})$ *for* $r \leq 1$, *while* $K^{-m}(X)$ *is naturally isomorphic to the subgroup* $\bigcap_{j=1}^{m} \ker q_j$ *of* $K^0(X\times S^{(m)})$.

If Θ_m^k and Δ_m^k denote the projection maps

$$\Theta_m^k : X \times S^{(k)} \to X \times S^{(k-m)}$$

$$\Delta_m^k : X \times S^{(k)} \to X \times S^{(m)}$$

obtained by deleting the last m circles and the first k–m circles, respectively; then for α in $K^{-m}(X)$ and a in $E_{1-k}(X)$ we define

$$\alpha \cap a = \Theta_{m*}^k(\Delta_m^{k*}(\alpha) \cdot a) \text{ in } E_{1-k+m}(X) ,$$

where the denotes module multiplication. This pairing is bilinear and associative with respect to the cap product in K-theory and natural in X in the appropriate way.

We now proceed to the definition of Per_* which holds no surprise. Recall that

$$K^{-2}(S^{(0)}) = \tilde{K}^{-2}(S^0) = \tilde{K}^0(S^2) = \mathbf{Z} ,$$

Moreover, if L is the Hopf bundle over S^2 (so L is a trivial line bundle on each hemisphere of S^2 with transition function defined at (x, y) on the equator to be multiplication by $x + iy$, then

$$\alpha_0 = [L] - [1]$$

is a generator for $K^{-2}(S^{(0)})$. The periodicity homomorphisms

$$\mathrm{Per}_* : \mathrm{Ext}_{r-2}(X) \to \mathrm{Ext}_r(X),\ r \leq 1$$

are defined by

$$\mathrm{Per}_*(a) = p^*(\alpha_0) \cap a ,$$

where $p : X \to S^{(0)}$. One of our most important results is

THEOREM 12. Per_* *is an isomorphism.*

The formal properties of Per_* follow from those of the cap product. In particular, Per_* is natural with respect to the long exact sequence.

Further, Per_* and Per^* are dual in the sense that the following diagram commutes:

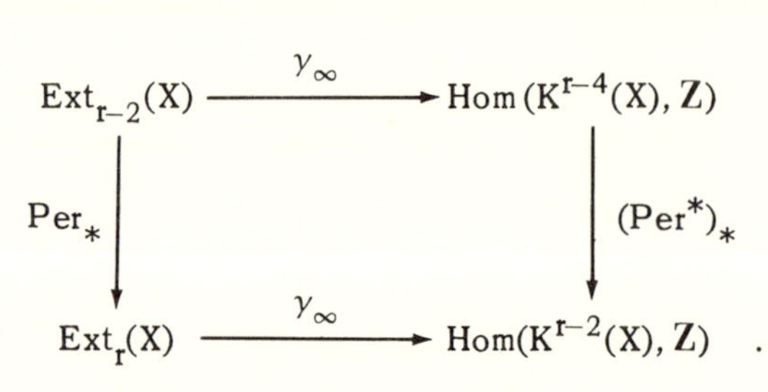

Note that this diagram suggests that we should have defined $\mathrm{Ext}(X) = \mathrm{Ext}_{-1}(X)$. This is Graeme Segal's contention [87] also from considerations of the real case where the period is 8. (Let me point out that although I have little doubt that it all works out, no one to my knowledge has checked the details for Ext in the real case; at least I have not!) Lastly, from the definition of the higher Ext groups, it follows that to prove Per_* is an isomorphism it is sufficient to consider the case $r = 1$ or $\mathrm{Per}_* : \mathrm{Ext}_{-1}(X) \to \mathrm{Ext}_1(X)$ for each X.

We will try to give some insight and to describe some of the techniques which go into the proof of periodicity. We begin with injectivity. As we said above, standard arguments show that $\mathrm{Ext}_{-1}(X) \equiv \mathrm{Ext}(S^2X)$ is naturally isomorphic to the subgroup of $\mathrm{Ext}(X \times S^2)$ of elements which map to 0 in both $\mathrm{Ext}(X)$ and $\mathrm{Ext}(S^2)$. (The second group is actually 0 and hence the latter assumption is unnecessary.) Let $p : X \times S^2 \to X$ be the projection. We want to show that for a in $\mathrm{Ext}(X \times S^2)$ such that $p_*(a) = 0$ and $\mathrm{Per}_*(a) = 0$, it follows that $a = 0$. If we let L denote the pullback of the Hopf bundle to $X \times S^2$, then one can show that

$$\mathrm{Per}_*(a) = p_*([L] \cdot a - a) .$$

Hence we want to show that for a in $\mathrm{Ext}(X \times S^2)$ satisfying $p_*(a) = p_*([L] \cdot a) = 0$, it follows that $a = 0$.

Let $\tau_0 : C(X \times S^2) \to \mathfrak{Q}(\mathfrak{H})$ define a. Since the inclusion of bundles $1 \subset 1 \oplus L$ induces an isomorphism of the Ext groups

$$\mathrm{Ext}(X\times S^2) = \mathrm{Ext}(C_1(X\times S^2)) \cong \mathrm{Ext}(C_{1\oplus L}(X\times S^2)),$$

we have a $\tau: C_{1\oplus L}(X\times S^2) \to \mathfrak{Q}(\mathfrak{H})\otimes M_2$ such that the decomposition

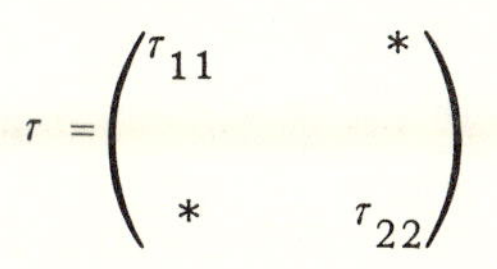

$$\tau = \begin{pmatrix} \tau_{11} & * \\ * & \tau_{22} \end{pmatrix}$$

has $[\tau_{11}] = a$ and $[\tau_{22}] = [L]\cdot a$. (We are using the fact that for any line bundle M over a space Y, we have $C_M(Y) = C(Y)$.) Now using off-diagonal elements t_1 and t_2 in τ which intertwine $\operatorname{im}\tau_{11}$ and $\operatorname{im}\tau_{22}$, one constructs the "polar decompositions" of t_1 and t_2 obtaining unitary elements u_1 and u_2 which satisfy $u_1\tau_{11}(f) = \tau_{22}(f)u_1$ so long as f vanishes on $X\times\{n\}$, where n denotes the north pole of S^2. Therefore a and $[L]\cdot a$ map to the same element in $\mathrm{Ext}(X\times S^2/X\times\{n\})$; hence there exists b in $\mathrm{Ext}(X)$ such that

$$i_*(b) = a-[L]\cdot a,$$

where i is the inclusion $X \to X\times\{n\} \subset X\times S^2$. But

$$b = p_*i_*(b) = p_*(a) - p_*([L]\cdot a) = 0$$

by hypothesis which implies $a = [L]\cdot a$.

Thus we can assume $\tau_{11} = \tau_{22} = \tau_0$ and use t_1 and t_2 to generate a larger commutative C^*-subalgebra of $\mathfrak{Q}(\mathfrak{H})$ with maximal ideal space Y. Thus we obtain c in $\mathrm{Ext}(Y)$ and a surjection $r: Y \to X\times S^2$ such that $r_*(c) = a$. One completes the proof of injectivity by studying the action of r. We need a concrete realization of the Hopf map. Let $S^{3'}$ be $I\times S^1\times S^1$ with the identifications $(t,\lambda,\mu) = (t',\lambda',\mu')$ if $t = t' = 0$ and $\lambda = \lambda'$ or $t = t' = 1$ and $\mu = \mu'$. Let $S^{2'}$ be $I\times S'$ with $(t,\lambda) = (t',\lambda')$ if $t = t' = 0$ or $t = t' = 1$. Define $h: S^{3'} \to S^{2'}$ by $h(t,\lambda,\mu) = (t,\lambda\mu)$. We show there is a map $\chi: Y \to X\times S^{2'}$ and a homeomorphism $\phi: S^2 \to S^{2'}$ such that the following diagram commutes

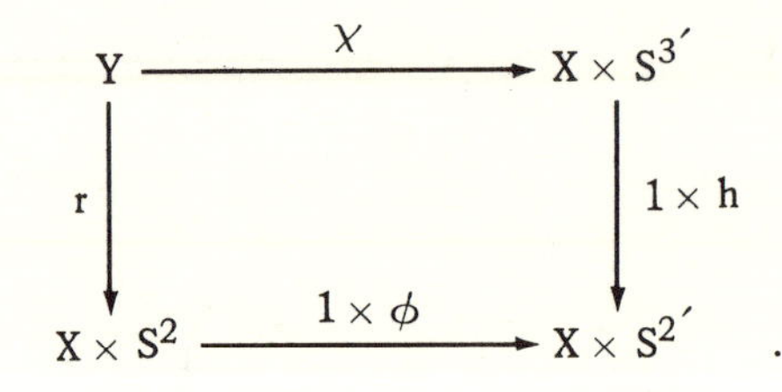

If p denotes the projection of all products onto X, then

$$p_* \chi_*(c) = p_*(1 \times h)_* \chi_*(c) = p_*(1 \times \phi)_*(a) = p_*(a) = 0 .$$

We now need

LEMMA. *If* d *in* $\mathrm{Ext}(X \times S^{2'})$ *projects to* 0 *in* $\mathrm{Ext}(X)$, *then*

$$(1 \times h)_*(d) = 0 .$$

Therefore $(1 \times \phi)_*(a) = (1 \times \phi)_* r_*(c) = (1 \times h)_* \chi_*(c) = 0$ and since $1 \times \phi$ is a homeomorphism we have $a = 0$ which shows that Per_* is injective.

The basic ingredient in proving the lemma is the following

LEMMA. *Let* Θ_1, Θ_2 *and* $\Theta_3 : X \times S^1 \times S^1 \to X \times S^1$ *be defined by* $(x, \lambda, \mu) \to (x, \lambda)$, (x, μ), $(x, \lambda\mu)$, *respectively. If* a *in* $\mathrm{Ext}(X \times S^1 \times S^1)$ *projects to* 0 *in* $\mathrm{Ext}(X)$, *then*

$$\Theta_{3*}(a) = \Theta_{1*}(a) + \Theta_{2*}(a) .$$

The proof depends on the familiar fact that $\begin{pmatrix} \lambda & 0 \\ 0 & \mu \end{pmatrix}$ and $\begin{pmatrix} \lambda\mu & 0 \\ 0 & 1 \end{pmatrix}$ are homotopic in U_2. This is the first hint that the group operation in $\mathrm{Ext}_0(X)$ can be defined using multiplication. We shall discuss this later.

COROLLARY. $\mathrm{Ext}(S^n) = 0$ *for* n *even.*

The proof of surjectivity involves the construction of an inverse using homotopy invariance. Let a in $\mathrm{Ext}(X)$ be defined by $\tau : C(X) \to \mathfrak{Q}(\mathfrak{H})$.

If $i_{\pm} : X \to X\times[-1,1]$ denote the two inclusion maps to $X \times \{\pm 1\}$, then $(i_+)_*(a) = (i_-)_*(a)$. Let $\tau_0 : C(X\times[-1,1]) \to \mathfrak{Q}(\mathfrak{H}_0)$ define the trivial extension and define

$$\tau_{\pm} = \tau(i_{\pm})^* \oplus \tau_0 : C(X\times[-1,1]) \to \mathfrak{Q}(\mathfrak{H}\oplus\mathfrak{H}_0) .$$

Then $[\tau_+] = [\tau_-]$ by homotopy invariance and hence there exists a unitary u in $\mathfrak{Q}(\mathfrak{H}\oplus\mathfrak{H}_0)$ such that $\tau_- = u\tau_+u^*$. If

$$j : C(X\times[-1,1]) \to C(X\times S^1) ,$$

then $\tau_+ = \tau_-$ on $j^*[C(X\times S^1)]$ and hence u commutes with the range of $\tau' = j_*(\tau_+) = j_*(\tau_-)$. We let b denote the element of $\operatorname{Ext}((X\times S^1)\times S^1)$ generated by $\operatorname{im}\tau'$ and u. It is easy to check that b projects to a in $\operatorname{Ext}(X)$. If q is the map from $S^1 \times S^1$ to S^2 obtained by collapsing $(S^1\times\{1\}) \cup (\{1\}\times S^1)$ to a point, then $c = (1\times q)_*(b)$ is an element of $\operatorname{Ext}(X\times S^2)$. If $p : X\times S^2 \to X$ is projection and $i : X \to X\times S^2$ is the inclusion obtained by choosing a base point, then for a in $\operatorname{Ext}(X\times S^2)$, $d - i_*p_*(d)$ projects to 0 in $\operatorname{Ext}(X)$. Moreover, since $\operatorname{Ext}(S^2) = 0$, we see that $d - i_*p(d)$ is in $\operatorname{Ext}_{-1}(X)' = \operatorname{Ext}(S^2X) \subset \operatorname{Ext}(X\times S^2)$. Therefore, $p_*([L]\cdot(d - i_*p_*(d))$ is in the range of Per_*. Applying this to c we have $p_*([L]\cdot c) - a$ is in the range of Per_*. The proof is now completed by showing that $p_*([L]\cdot c) = 2a$ by using a calculation which we omit.

If $p : E \to X$ is an $(n+1)$-dimensional real vector bundle over X, then E is said to be *oriented* for K-theory if there exists α in $K^n(S(E))$, where $S(E)$ denotes the S^n-bundle in E over X and $p' : S(E) \to X$, which restricts to a generator of $\tilde{K}^n(p'^{-1}(x))$ for each x in X. The element α is said to be a *K-theory orientation* for E and is not unique, in general. Using standard arguments it is possible to establish a Thom isomorphism theorem ([6], p. 78).

THEOREM 13. *If* $p : E \to X$ *is a rank* $n+1$ *vector bundle over* X *and* α *in* $K^n(S(E))$ *is a K-theory orientation, then the map*

$$\tau_q : K_q \rightarrow E_{q-n}(X)$$

defined by $\tau_q(a) = p'_*(\alpha \cap a)$ *is an isomorphism, where* K_q *is the kernel of* $p_* : \mathrm{Ext}_q(S(E)) \to \mathrm{Ext}_q(X)$.

The proof proceeds in three steps. First, one considers the case of a product $X \times R^{n+1}$, where α comes from a generator of $\tilde{K}^n(S^n)$, which follows since τ_q is a power of Per_*. Second, one allows α to be the product of a generator by a unit of $K^0(X)$. Finally, one performs induction on the number of open sets necessary to trivialize E, using the Mayer-Vietoris sequence and the naturality of τ_q.

CHAPTER 5
Ext AS K-HOMOLOGY

With the periodicity theorem in hand, a reduced generalized homology theory Ext_* is obtained by defining

$$Ext_n(X) = \begin{cases} Ext(X) & n \text{ odd} \\ Ext(SX) & n \text{ even} \end{cases}$$

and a non-reduced theory E_* by defining

$$E_n(X) = \begin{cases} Ext(X) & n \text{ odd} \\ \ker \pi_* & n \text{ even} \end{cases}$$

where $\pi: X \times S^1 \to X$. Then Ext_* and E_* satisfy the Eilenberg-Steenrod axioms except for the dimension axiom. That is, $X \mapsto Ext_n(X)$ is a sequence of covariant, homotopy invariant functors with boundary maps $\partial: Ext_n(X/A) \to Ext_{n-1}(A)$ for $A \subset X$ such that the following sequence

$$\cdots \xrightarrow{\partial} Ext_n(A) \xrightarrow{i_*} Ext_n(X) \xrightarrow{p^*} Ext_n(X/A) \xrightarrow{\partial} Ext_{n-1}(A) \longrightarrow \cdots$$

is exact, where $i: A \to X$ is inclusion and $p: X \to X/A$ the quotient map. The dimension axiom is replaced by

COROLLARY. $Ext_q(S^0) = Ext(S^{q-1}) = \begin{cases} Z & q \text{ even} \\ 0 & q \text{ odd} \end{cases}$.

If $\mathcal{T}_{B^n}$ denotes the Toeplitz extension defined in the first chapter for B^n the unit ball in C^n, then $\mathcal{T}_{B^n}$ is in $Ext(S^{2n-1})$ since $\partial B^n = S^{2n-1}$.

By the index results of Venugopalkrishna [97] or Boutet de Monvel [13], it follows that $\mathcal{T}_{B^n}$ is a generator for $\mathrm{Ext}(S^{2n-1})$ and hence these extensions form the "cells" for $\mathrm{Ext}(X)$ for X a finite complex.

Now Ext_* actually satisfies the two axioms introduced by Milnor [58] for the purpose of characterizing the Steenrod homology theory [94] for compact metrizable spaces. These are (1) the strong excision axiom and (2) a "cluster axiom". A homology theory has the *strong excision property* if for each compact space X, closed subset A of X, and open subset U of X which is contained in A, it follows that inclusion

$$i : (X \setminus U, A \setminus U) \to (X, A)$$

induces an isomorphism

$$i_* : \mathrm{Ext}_n(X \setminus U, A \setminus U) \to \mathrm{Ext}_n(X, A) \text{ for all } n .$$

This is automatic for Ext_* since the relative groups are defined by

$$\mathrm{Ext}_*(X, A) = \mathrm{Ext}_*(X/A) .$$

A compact metrizable space X is said to be the *strong wedge* of the closed subspaces $X_1, X_2 \cdots$ of X if

$$X = \cup X_n, \text{ diam } X_n \to 0 \text{ and } X_n \cap X_m = \{b\} \text{ for } n \neq m$$

for some point b in X.

THEOREM 14. *If X is the strong wedge of the closed subspaces $X_1, X_2 \cdots$ and $r_n : X \to X_n$ is the retraction carrying X_j to $\{b\}$ for $j \neq n$, then*

$$a \to (r_{1*}(a), r_{2*}(a), \cdots)$$

defines an isomorphism of $\mathrm{Ext}_q(X)$ onto $\prod_{n=1}^{\infty} \mathrm{Ext}_q(X_n)$ for all q.

The proof depends on the fact that one can define an infinite sum $\Sigma \tau_n$ so long as $\|\tau_n(f)\| \to 0$ for each f in $C(X)$ which is the case here since diam $X_n \to 0$. In general, one can not identify the infinite direct sum $\mathfrak{Q}(\mathfrak{H}) \oplus \mathfrak{Q}(\mathfrak{H}) \oplus \cdots$ with $\mathfrak{Q}(\mathfrak{H} \oplus \mathfrak{H} \oplus \cdots)$ because the direct sum $K_1 \oplus K_2 \oplus \cdots$ for K_i compact is not always compact. It is compact if and only if $\|K_i\| \to 0$. Let $\{e_i\}$ be the projections in $\mathfrak{Q}(\mathfrak{H} \oplus \mathfrak{H} \oplus \cdots)$ corresponding to the summands. For f in C(X) such that $f(b) = 0$, then define

$$\tau(f) = \sum_{i=1}^{\infty} e_i \tau_i(f \mid X_i) e_i$$

which converges since $\|f \mid X_i\| \to f(b) = 0$.

An example of a strong wedge is the so-called Hawaiian earring W

which is the strong wedge of a countable family of circles for which Ext is the direct product of infinitely many copies of $\mathbf{Z}$. This follows since W is a subset of $\mathbf{C}$ and $\operatorname{ind}(T-\lambda I)$ may be chosen arbitrarily for any component of $\mathbf{C} \setminus W$; thus $\Pi \mathbf{Z}$ is appropriate and not $\oplus \mathbf{Z}$.

An important consequence of the fact that Ext_* is a Steenrod homology theory is the fact that the kernel of $P: \mathrm{Ext}(\varprojlim X_n) \to \varprojlim \mathrm{Ext}(X_n)$ can be determined using the first derived functor $\varprojlim^{(1)}$ for inverse limit. Recall that $\lim^{(1)} G_n$ of an inverse system $\{G_n, \phi_n\}$ of groups and homomorphisms is the cokernel of the homomorphism $\Pi G_n \to \Pi G_n$ defined by

$$(g_1, g_2, \cdots) \to (g_1 - \phi_2(g_2), g_2 - \phi_3(g_3), \cdots).$$

There is an exact sequence

$$0 \longrightarrow \varprojlim^{(1)}(\mathrm{Ext}_{k+1}(X_n)) \longrightarrow \mathrm{Ext}_k(\varprojlim X_n) \xrightarrow{P} \varprojlim(\mathrm{Ext}_k(X_n)) \longrightarrow 0$$

which measures the lack of continuity of Ext with respect to inverse limit. This plus techniques from algebraic K-theory are the ingredients of L. G. Brown's proof of the universal coefficient theorem [18] which states that the sequence

$$0 \longrightarrow \mathrm{ext}(K^0(X), Z) \longrightarrow \mathrm{Ext}(X) \xrightarrow{\gamma_\infty} \mathrm{Hom}(K^1(X), Z) \longrightarrow 0$$

is exact and splits but non-canonically. The "ext-group" denotes the usual group consisting of the group extensions of $K^0(X)$ by Z. (After Ext_* and K_* are identified for finite complexes, this result is seen to extend the universal coefficient theorem of D. Anderson [2] for K-theory.)

These properties enable Ext of many standard spaces to be calculated. For example using Atiyah's calculations in K-theory [6] along with the universal coefficient theorem the following can be established [48]: If X is complex projective space or a complex Grassmannian, then $\mathrm{Ext}_n(X) = 0$ for n odd and $\mathrm{Ext}_n(X) = Z^{r-1}$ for n even, where r is the number of even-dimensional cells required where X is constructed using only even-dimensional cells in a minimal way. Further, $\mathrm{Ext}_n(RP^k) = 0$ for n even while $\mathrm{Ext}_n(RP^{2k}) = Z_{2^k}$ and $\mathrm{Ext}_n(RP^{2k+1}) = Z \oplus Z_{2^k}$ for n odd. Also, $\mathrm{Ext}_n(U(k)) = Z^{2^{k-1}-1}$ for n even while $\mathrm{Ext}_n(U(k)) = Z^{2^{k-1}}$ for n odd. Finally, since Atiyah (for $U(n)$) and Hodgkin [44] (for G a compact, connected, simply-connected Lie group) have shown that $K^*(G)$ is an exterior algebra on certain representations of G, it follows that $\mathrm{Ext}_*(G)$ is free abelian. Recently, Cass and Snaith have explicitly constructed [24] a set of generators for G a compact, connected Lie group with $\pi_1(G)$ torsion-free.

Although we do not prove the universal coefficient theorem we do want to indicate how an element in the kernel of γ_∞ gives rise to a group extension. Recall that for Banach algebra $\mathcal{B}$ one defines the algebraic K-groups (cf. [50], [59], [96]) $K_0(\mathcal{B})$ and $K_1(\mathcal{B})$ as follows. Let $L_n(\mathcal{B})$

denote the quotient of $GL_n(\mathcal{B})$ by connected component of the identity and $L_n(\mathcal{B}) \to L_{n+1}(\mathcal{B})$ the natural map induced by the map $B \to B \oplus 1$ defined from $M_n(\mathcal{B}) \to M_{n+1}(\mathcal{B})$. We define $K_1(\mathcal{B}) = \varinjlim L_n(\mathcal{B})$. For $\mathcal{B} = C(X)$ we have $K_1(C(X)) = K^1(X)$. Further, let $\mathcal{P}_n(\mathcal{B})$ denote the idempotents in $M_n(\mathcal{B})$. For A in $\mathcal{P}_m(\mathcal{B})$ and B in $\mathcal{P}_n(\mathcal{B})$ we let $A \sim B$ if there exists k such that $A \oplus 0$ and $B \oplus 0$ in $\mathcal{P}_{m+n+k}(\mathcal{B})$ are similar and we let $\mathcal{T}(\mathcal{B})$ denote the collection of equivalence classes in $\cup \mathcal{P}_n(\mathcal{B})$. An addition is defined on $\mathcal{T}(\mathcal{B})$ such that $[A] + [B] = [A \oplus B]$ and we let $\mathcal{U}\mathcal{T}(\mathcal{B})$ denote the universal group constructed from $\mathcal{T}(\mathcal{B})$. Lastly, we let $K_0(\mathcal{B})$ denote $\mathcal{U}\mathcal{T}(\mathcal{B})$ modulo the subgroup generated by elements of the form $[A+B] - [A] - [B]$ where A and B are commuting disjoint elements of $\mathcal{P}_n(\mathcal{B})$. Again $K_0(C(X)) = K^0(X)$.

Let $(\mathfrak{S}, \phi)$ be an element of $\mathrm{Ext}(X)$ such that $\gamma_\infty(\mathfrak{S}, \phi) = 0$. Then corresponding to the sequence $0 \to \mathcal{K} \to \mathfrak{S} \overset{\phi}{\to} C(X) \to 0$ is the exact sequence in algebraic K-theory (cf. [59])

$$K_1(\mathfrak{S}) \longrightarrow K_1(C(X)) \overset{\partial}{\longrightarrow} K_0(\mathcal{K}) \longrightarrow K_0(\mathfrak{S}) \longrightarrow K_0(C(X)) \longrightarrow 0 .$$

An invertible element t in $\mathcal{Q}(\mathfrak{H})$ lies in $\pi(\mathcal{L}(\mathfrak{H})^{-1})$ if and only if $\mathrm{ind}(T) = 0$ for T in $\mathcal{L}(\mathfrak{H})$ such that $\pi(T) = t$. Therefore, since $\gamma_\infty(\mathfrak{S}, \phi) = 0$ it follows that each invertible element of $C(X) \otimes M_n$ lifts to an invertible element of $\mathfrak{S} \otimes M_n$. Therefore the map $K_1(\mathfrak{S}) \to K_1(C(X))$ is onto and since $K_0(\mathcal{K}) = \mathbf{Z}$, we obtain

$$0 \to \mathbf{Z} \to K_0(\mathfrak{S}) \to K^0(X) \to 0$$

which is the desired element of $\mathrm{ext}(K^0(X), \mathbf{Z})$. One of course must show that every element arises in this manner and that this extension determines $(\mathfrak{S}, \phi)$.

That the universal coefficient sequence splits follows from the facts that $\mathrm{ext}(K^0(X), \mathbf{Z})$ is algebraically compact, while $\mathrm{Hom}(K^1(X), \mathbf{Z})$ is torsion-free. From this splitting, and Anderson's universal coefficient theorem for topological K-theory [2] it follows that $E_*(X)$ and $K_*(X)$ are abstractly isomorphic for X a finite complex, where one defines K_* using Spanier-Whitehead duality [93]. For that definition one embeds X as a closed subset of S^{2n+1}, chooses a strong deformation retract $D_{2n}(X)$ of $S^{2n+1} \setminus X$ and then defines

$$K_*(X) = K^*(D_{2n}(X)) .$$

This is well defined since $D_{2m}(X)$ and $D_{2n}(X)$ will have the same stable homotopy type. It is tempting to attempt to define $K_*(X) = K^*(S^{2n+1} \setminus X)$ for $X \subset S^{2n+1}$. Although the two are isomorphic by Steenrod duality [46], such a definition is not very useful since $X \mapsto K^*(S^{2n+1} \setminus X)$ does not define a functor.

Now although such a definition yields K_* for finite complexes, it is useful to have a concrete definition. In [7] Atiyah attempted such a definition by generalizing the notion of elliptic operator. For X a compact metrizable space let $\sigma_1 : C(X) \to \mathcal{L}(\mathfrak{H}_1)$ and $\sigma_2 : C(X) \to \mathcal{L}(\mathfrak{H}_2)$ be $*$-homomorphisms. An operator $T : \mathfrak{H}_1 \to \mathfrak{H}_2$ is said to be *elliptic* relative to (σ_1, σ_2) if

1) T is Fredholm

and

2) $T\sigma_1(f) - \sigma_2(f)T$ is compact for f in $C(X)$.

Atiyah then let $\mathrm{Ell}(X)$ denote the collection of triples (σ_1, σ_2, T) with addition defined by direct sum, that is, $(\sigma_1, \sigma_2, S) + (\tau_1, \tau_2, T) = (\sigma_1 \oplus \tau_1, \sigma_2 \oplus \tau_2, S \oplus T)$. The classic example is obtained by letting σ_1, and σ_2 be the representations defined by multiplication on the L^2 space of a smooth measure on a smooth manifold X, and where T is a pseudo-differential operator of order zero.

Atiyah defined a pairing $\mathrm{Ell}(X) \times K^0(X) \to \mathrm{Ell}(X)$ for X a finite complex as follows: Let (σ_1, σ_2, T) be an element of $\mathrm{Ell}(X)$ and V be a vector bundle on X. Since there exists a complementary vector bundle W such that $V \oplus W$ is isomorphic to the trivial bundle $X \times \mathbf{C}^n$ for some n, we can find a projection valued function P from X to $M_n(\mathbf{C})$ such that the fiber of W at x is equal to the kernel of $P(x)$. The operator $P_i = (\sigma_i \otimes 1_n)(P)$ is a projection on $\mathfrak{H}_i \otimes C^n$ which commutes with $\sigma_i \otimes 1_n$ for $i = 1, 2$. Moreover, P_2TP_1 defines an operator from $P_1(\mathfrak{H}_1 \otimes C^n$ to $P_2(\mathfrak{H}_2 \otimes C^n)$ such that the triple $(P_1(\sigma_1 \otimes 1_n)P_1, P_2(\sigma_2 \otimes 1_n)P_2, P_2TP_1)$ defines the element of $\mathrm{Ell}(X)$ which is the product of $(\sigma_1, \sigma_2, \tau)$ by $[V]$. One must show, of course, that this product is well defined. Next one generalizes this construction to the case where V is a vector bundle over $X \times Y$ to obtain the element of $K^0(Y)$ defined by the continuous map $y \to P_2(y)TP_1(y)$ from Y to the Fredholm operators. Although these Fredholm operators are defined on a Hilbert bundle, this causes no problems since all such bundles are trivial [31]. Thus one obtains the product $\mathrm{Ell}(X) \times K^0(X \times Y) \to K^0(Y)$. This can be viewed as the ingredients for a "slant" product pairing between a homology theory and its dual cohomology theory [92].

Atiyah then used this slant product to define a map $\mathrm{Ell}(X) \to K_0(X)$, where $K_0(X)$ is again defined by Spanier-Whitehead duality using D_nX. Indeed, if μ is a fixed fundamental element in $K^0(X \times D_nX)$, then for (σ_1, σ_2, T) in $\mathrm{Ell}(X)$ we obtain $(\sigma_1, \sigma_2, T)/\mu$ which lies in $K^0(D_nX) = K_0(X)$ and hence we have the desired map $\mathrm{Ell}(X) \to K_0(X)$. If T is an elliptic pseudo-differential operator, then its image under this map defines an "analytic index" in $K_0(X)$. On the other hand, one obtains, in a standard way, a "topological index" in $K_0(X)$. Using the index theorem for families Atiyah extended the index theorem to show that the topological and analytical index agree as elements of $K_0(X)$. Further, by appealing to a result of Conner and Floyd, he shows that the map is onto $K_0(X)$ for finite complexes and then states the problem of determining the equivalence relation on $\mathrm{Ell}(X)$ necessary to define $K_0(X)$. Our work provides a solution.

Let (σ_1, σ_2, T) be an element of $\mathrm{Ell}(X)$ for which $\sigma_1 = \sigma_2 = \sigma$. (One can show that such elements generate $K_0(X)$.) Further, let $\tau = \pi \circ \sigma$ and $\pi(T) = uh$ be the polar decomposition of $\pi(T)$, which is possible since $\pi(T)$ is invertible. One sets $h = \sqrt{\pi(T)^* \pi(T)}$ and $u = \pi(T)h^{-1}$, where the latter element is unitary. If $\mathfrak{Z}$ denotes the C*-subalgebra of $\mathfrak{Q}$ generated by $\operatorname{im} \tau$ and u, then $\mathfrak{Z} \cong C(Y)$, where $Y \subset X \times S^1$. Therefore, we obtain $\tau' : C(X \times S^1) \to \mathfrak{Q}$ such that $p_*([\tau']) = [\tau] = 0$, and hence an element $[\tau']$ in $E_0(X)$. Thus we obtain a map $\mathrm{Ell}(X) \to E_0(X)$.

Using our slant product which is basically the same as Atiyah's we obtain, also using Atiyah's argument for the existence of a map but now for Ext, the diagram

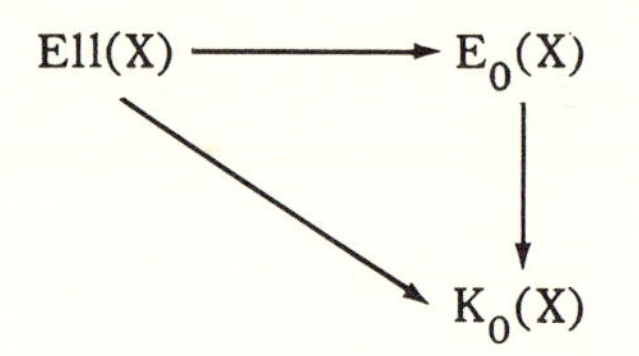

Then a map $E_{-1}(X) \to K_{-1}(X)$ is defined and after checking that the periodicity maps match up, a map $E_* \to K_*$ between homology theories is obtained. Moreover, since this map is a natural transformation which is an isomorphism on spheres, it follows that it is an isomorphism on finite complexes.

Now in addition to identifying E_* and K_* for finite complexes, this diagram solves Atiyah's problem. For (σ', T') and (σ'', T'') in $\mathrm{Ell}(X)$, we have $(\sigma', T') \sim (\sigma'', T'')$ if and only if there exists a *-homomorphism $\sigma_0 : C(X) \to \mathfrak{L}(\mathfrak{H}_0)$, a unitary U_0 on $\mathfrak{H}_0$ commuting with $\operatorname{im} \sigma_0$ and a unitary V from $\mathfrak{H}' \oplus \mathfrak{H}_0$ to $\mathfrak{H}'' \oplus \mathfrak{H}_0$ such that

$$V(\sigma'(f) \oplus \sigma_0(f))V^* - \sigma''(f) \oplus \sigma_0(f) \text{ is compact for } f \text{ in } C(X)$$

and

$$\pi(V)(u' \oplus \pi(u_0))\pi(V)^* = u'' \oplus \pi(u_0),$$

where u' and u'' are the unitary factors of the polar decompositions for $\pi(T')$ and $\pi(T'')$, respectively. Instead of involving the polar decomposition, one could allow an equivalence defined by a homotopy of the operator T keeping σ fixed.

In [46] Kahn, Kaminker and Schochet show that for any cohomology theory constructed from a spectrum, it is possible to define an associated Steenrod homology theory in a concrete fashion on the category of compact metrizable spaces. Since K-theory can be defined using a spectrum, there is a Steenrod homology theory associated which they show is naturally equivalent to E_* on the category on finite dimensional compact metrizable spaces. As corollaries they obtain the universal coefficient theorem and strong homotopy invariance for E_* (which we will explain presently) for finite dimensional spaces. In addition they establish the Steenrod duality theorem: for X a closed subspace of $\mathbf{C}^n$ $\mathrm{Ext}(X) \cong \tilde{K}^0(\mathbf{C}^n \setminus X)$, which generalizes the theorem stated in Chapter 1 for $X \subset \mathbf{C}$.

There is another solution due to Kasparov [52], [53] to Atiyah's problem of characterizing the equivalence relation on $\mathrm{Ell}(X)$ defined by $K_*(X)$. Basically Kasparov defines a contravariant functor on the category of complex involutive, finitely generated Banach algebras allowing a compact group of automorphisms as well as a Clifford algebra to act. In his definition the existence of inverses is built in and his notion of equivalence includes homotopy. In general he obtains a periodic generalized homology theory on finite-dimensional compact metrizable spaces. In the context of the semigroup $\mathrm{Ell}(X)$, he shows that $K_0(X)$ is obtained by taking the smallest equivalence relation generated by

1) $(\sigma, T) \sim (U\sigma^*U^*, UTU^*)$ for U unitary,

2) $(\sigma, T) \sim 0$ if T actually commutes with $\mathrm{im}\,\sigma$ and $\mathrm{ind}\, T = 0$, and

3) $(\sigma, T_0) \sim (\sigma, T_1)$ if (σ, T_λ) is a homotopy in $\mathrm{Ell}(X)$.

While in our work the basic operation is the module action, Kasparov constructs a product

$$K_n(X) \times K_m(Y) \to K_{n+m}(X \times Y)$$

which is fundamental. The difficulty in defining this product is the usual one; namely, the tensor product $S \otimes T$ of Fredholm operators S and T is not Fredholm. Kasparov substitutes a weighted average of $S \otimes I$ and

$I \otimes T$ which he shows is unique up to equivalence. There are many other interesting results and the complete relation of Kasparov's work to ours is yet to be determined.

Using our results and Kasparov's one can argue that strong homotopy invariance holds for Ext; that is, if τ_0 and τ_1 are *-monomorphisms from $C(X)$ into $\mathfrak{Q}(\mathfrak{H})$ which are homotopic, then $[\tau_0] = [\tau_1]$ in $\text{Ext}(X)$. Since index invariants are obviously preserved under homotopy, the problem can be reduced to elements in the kernel of γ_∞. L. G. Brown has also given a proof of strong homotopy invariance by using techniques from algebraic K-theory. As mentioned earlier the results of Kahn, Kaminker, and Schochet establish it for finite dimensional compact metrizable spaces.

These ideas also allow one to give an alternate definition of addition in $E_0(X)$. Recall that a pair of elements of $E_0(X)$ can be represented by a *-monomorphism $\sigma: C(X) \to \mathfrak{Q}(\mathfrak{H})$ and essential unitaries u_0 and u_1 in $\mathfrak{Q}(\mathfrak{H})$. Then one can show that $[(\sigma, u_0)] + [(\sigma, u_1)] = [(\sigma, u_0 u_1)]$.

There is a direct relation between Ext and K-homology which was suggested by Atiyah and Segal [8]. Let X be a compact subset of R^{2n} and $\tau: C(X) \to \mathfrak{Q}(\mathfrak{H})$ define an element of $\text{Ext}(X)$. Choose $A_1, A_2, \cdots A_{2n}$ in $\mathfrak{L}(\mathfrak{H})$ such that $\pi(A_i) = \tau(x_i)$, the coordinate functions, and let $E_1, E_2, \cdots E_{2n}$ be the Clifford matrices on $C^{2^{n-1}}$ which satisfy $E_i^2 = I$ and $E_i E_j = -E_j E_i$, $i \neq j$. Define $F: R^{2n} \to \mathfrak{L}(\mathfrak{H}) \otimes M_{2^{n-1}}(C)$ by

$$F(\mu) = \sum_{i=1}^{2n} (A_i - \mu_i) \otimes E_i \,.$$

Since $F(\mu)^2 = \sum_{i=1}^{2n} (A_i - \mu_i)^2 \otimes I$ modulo $\mathcal{K}$, we see that $F(\mu)$ is Fredholm if and only if μ is not in X. Therefore, we have $F: R^{2n} \setminus X \to \text{Fred}(H \otimes C^{2^{n-1}})$ or an element of $K^0(R^{2n} \setminus X)$ which equals $K_1(X)$ by Spanier-Whitehead duality.

The Atiyah-Segal map does yield an isomorphism $\text{Ext}(X) \cong K^0(R^{2n} \setminus X)$, but in order to prove that one must use our machinery. In [46] Kahn, Kaminker, and Schochet show that their isomorphism

$$\operatorname{Ext}(X) \cong K_1(X) \cong \tilde{K}^0(R^{2n} \setminus X),$$

where the first map is constructed using our machinery while the second comes from duality and corresponds to the Atiyah-Segal map.

Finally, there have been several other concrete realizations of K-homology defined recently by Sullivan [95], Morgan, and Baum [11]. Working out the precise relation between these theories, Kasparov's theory, and Ext, would seem to hold exciting possibilities.

CHAPTER 6

INDEX THEOREMS AND NOVIKOV'S HIGHER SIGNATURES

In this final chapter we wish to discuss the connection between Ext and various problems in algebraic topology. We begin by discussing a very general index theorem in the context of Ext first formulated by Singer [89].

If X is a compact space, then the Chern character defines a homomorphism $ch^*: K^1(X) \to H^{odd}(X, \mathbf{Q})$. Moreover, dually we have the Chern character for homology, which yields the homomorphism $ch_*: \mathrm{Ext}_1(X) \to H_{odd}(X, \mathbf{Q})$ (cf. [48]). Further, the pairing of Ext with K-theory matches up with the Kronecker pairing of homology and cohomology to yield the commutative diagram

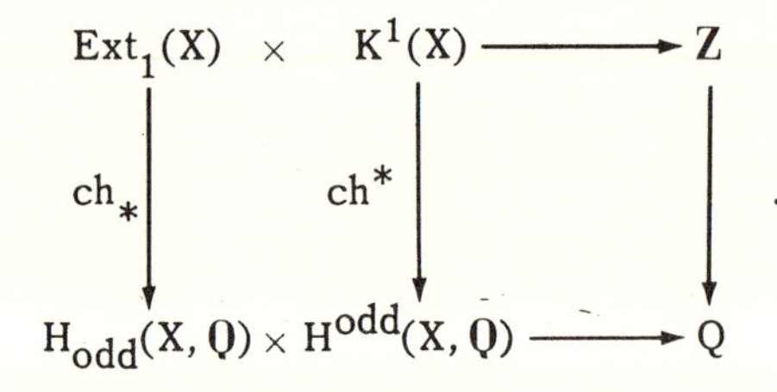

$$\begin{array}{ccccc} \mathrm{Ext}_1(X) & \times & K^1(X) & \longrightarrow & \mathbf{Z} \\ \downarrow ch_* & & \downarrow ch^* & & \downarrow \\ H_{odd}(X, \mathbf{Q}) & \times & H^{odd}(X, \mathbf{Q}) & \longrightarrow & \mathbf{Q} \end{array} .$$

Thus for $(\mathfrak{E}, \phi)$ in $\mathrm{Ext}(X)$ we have the "index class" $ch_*(\mathfrak{E}, \phi)$ in $H_{odd}(X, \mathbf{Q})$ for which the following index formula holds:

$$\mathrm{ind}(T) = \gamma_\infty(\mathfrak{E}, \phi)(\pi(T)) = \langle ch_*(\mathfrak{E}, \phi), ch^*(\pi(T)) \rangle ,$$

for T in $\mathfrak{E} \otimes M_k(\mathbf{C})$, where $\langle , \rangle$ denotes the natural pairing between homology and cohomology. This is as far as we can go in this generality. The problem is to give an explicit formula for $ch_*(\mathfrak{E}, \phi)$, when $(\mathfrak{E}, \phi)$ is given explicitly. If X is an oriented connected compact n-dimensional manifold, then we can reformulate the problem using Poincaré duality $D: H^i(X, Q) \to H_{n-i}(X, Q)$ to obtain

SINGER INDEX THEOREM. *If* X *is an oriented connected compact manifold and* $\mathcal{I}(\mathcal{E}, \phi) = D^{-1}(ch_*(\mathcal{E}, \phi))$ *is the index class in* $H^*(X, Q)$, *then*

$$\mathrm{ind}(T) = [\mathcal{I}(\mathcal{E}, \phi) \cup ch^*(\pi(T))][X]$$

for T *in* $\mathcal{E} \otimes M_k(C)$.

To make this theorem useful we must calculate $\mathcal{I}(\mathcal{E}, \phi)$ concretely for explicit extensions. For example, consider the extension $\mathcal{P}_M$ constructed from the algebra of zero-th order pseudo-differential operators on the closed manifold M. Then the index theorem of Atiyah and Singer [10] identifies $\mathcal{I}(\mathcal{P}_M)$ as $\tau(T(S^*(M)) \oplus \Theta^k)$, the Todd class of the stable tangent bundle of S^*M. Recall, that when a vector bundle admits a complex structure then the Todd class is a certain universal polynomial in the Chern classes of the bundle [10, III].

Also, the result of Venugopalkrishna [97] yields that $\mathcal{I}(\mathcal{T}_{B_n}) = 1$ which is the same as the Todd class determined by the complex structure. More recently, Boutet de Monvel has shown [13] in considerable generality, including all strongly pseudo convex subdomains of a Stein manifold, that $\mathcal{I}(\mathcal{T}_\Omega)$ is given by the Todd class determined by the complex structure. The problem of determining the index class for canonical extensions defined on contact manifolds has been raised by Singer [89]. Dynin has announced results on this problem.

Now let us consider more carefully the element $[\mathcal{P}_M]$ in $\mathrm{Ext}_1(S^*(M))$ determined by the pseudo-differential operators of order zero on a smooth manifold M. First one can show that $[\mathcal{P}_M]$ is a K-theory fundamental class for $S^*(M)$, that is, for any open disk $D^{2n+1} \subset S^*(M)$ and the collapsing map $p: S^*(M) \to S^*(M)/(S^*(M) \setminus D^{2n+1}) \equiv S^{2n+1}$, $p_*[\mathcal{P}_M]$ is a generator in $\mathrm{Ext}_{2n+1}(S^{2n+1}) = Z$. One proof of this uses the Atiyah-Singer index theorem to show that $ch_*[\mathcal{P}_M] = \tau(S^*(M)) \cap [S^*(M)]$. A detailed exposition of this will form the preliminaries in a forthcoming paper by Kaminker. As a corollary to this one obtains the fact that the pseudo-differential algebra extension $[\mathcal{P}_M]$ is "locally isomorphic" to the Toeplitz operator extension in S^{2n+1}. A direct proof of this fact is given by Helton and Howe in [42] for the n-dimensional torus. A more refined version of this showing $[\mathcal{P}_M]$ and $[\mathcal{T}_\Omega]$ locally isomorphic for M a compact smooth manifold and Ω a strongly pseudo-convex domain is used by Boutet de Monvel [13] in his calculation of the index class of $\mathcal{T}_\Omega$. This enables him to use the Atiyah-Singer index theorem in the Toeplitz context. A more incisive proof, if possible, of these latter properties would allow one to show directly that these naturally occurring extensions on contact manifolds have the Todd class as their index class. Moreover, it would extend the class of the examples for which one had an explicit index theorem. This problem has been raised by Singer [89].

The correspondence $M \mapsto [\mathcal{P}_M]$ defines a mapping from manifolds to K-theory fundamental classes. The determination of exactly which class one has is extremely important. If this correspondence was well behaved functorially, one could attempt to determine which class one had by some sort of inductive technique. The fact that the class lies in $Ext_1(S^*(M))$ makes that extremely difficult. One would much prefer a K-theory fundamental class in $E_0(M)$. (There is a prime two torsion obstruction to the existence of a K-theory fundamental class on a manifold. Hence we assume for the rest of this paragraph that M is stable almost complex.) If $B^*(M)$ denotes the unit ball bundle contained in $T^*(M)$ and $Th^*(M)$, the Thom space which is the quotient $B^*(M)/S^*(M)$, then we have the sequence

$$E_0(Th^*(M)) \xrightarrow{\partial} E_1(S^*(M)) \xrightarrow{i_*} E_1(B^*(M)) .$$

Since $E_1(B^*(M)) = E_1(M)$ and $i_*[\mathcal{P}_M]$ restricts to the multiplication operators on $L^2(M)$, we have $i_*[\mathcal{P}_M] = 0$. Therefore, there exists a in

$E_0(Th^*(M))$ such that $\partial a = [\mathcal{P}_M]$. Now this element is unique if and only if $j_* = 0$, where $j_*: E_0(B^*(M)) \to E_0(Th^*(M))$. Now j_* can be identified with multiplication by the K-theory Euler class and is zero, for example, if the ordinary Euler class is zero. This follows from the existence of a section $s: M \to S^*(M)$ and since $i_* \circ s_* = 1$, we see that i_* is onto and hence $j_* = 0$. On the other hand the K-theory Euler class is not zero for CP^n. However, using additional structure (such as the stable almost complex structure on $Th^*(M)$), one may be able to choose a canonical a. Using the Thom isomorphism $E_0(Th^*(M)) \to E_0(M)$, we might obtain a fundamental class in $E_0(M)$. The difficulty in doing this canonically involves prime two torsion and in some attempts there is a noncanonical dependence on the smooth structure of M. This phenomena, however, allows one in some cases to use $[\mathcal{P}_M]$ to detect distinct smooth structures on M.

There is another approach to this difficulty. In general, the cosphere bundle $S^*(M)$ on a smooth manifold M depends on the smooth structure on M. However, one knows that $S^*(M)$ is naturally unique up to homotopy type and thus $\operatorname{Ext}(S^*(M))$ is determined by the topological type of M. Now the element $\gamma_\infty[\mathcal{P}_M]$ is determined by the Atiyah-Singer index theorem and is, in fact, independent of the smooth structure on M. This latter statement is equivalent to the theorem of Novikov on the topological invariance of the rational Pontryagin classes. One direction depends on the fact that the Todd class does not depend on the smooth structure, while the other depends on the fact that the rational Pontryagin classes can be determined by the index of operators in $\mathcal{P}_M$.

A different approach to proving the Novikov theorem might be based on $[\mathcal{P}_M]$. In particular, one sees that the Novikov theorem is equivalent to the fact that $[\mathcal{P}_{M_1}] - [\mathcal{P}_{M_2}]$ is a torsion element where M_1 and M_2 denote homeomorphic smooth manifolds. This is very close to Singer's program [88] to prove the Novikov theorem in which he attempts to define explicitly an analogue of $\mathcal{P}_M$ for M a piecewise linear manifold. Unfortunately, I can report no progress along these lines. Kasparov, however, claims [53] some success with this approach.

Recall that Kasparov defines his K-functors for arbitrary Banach algebras with involution and establishes good properties for his K-homology at least for Banach algebras which are finitely generated. Thus for π a finitely generated group we can consider the group algebra $\ell^1(\pi)$ or its enveloping C^*-algebra $C^*(\pi)$. And Kasparov defines a ring $R_*(\pi) = R_0(\pi) \oplus R_1(\pi)$ with $R_i(\pi) = K_i(C^*(\pi))$, $i = 0, 1$, where the ring multiplication is obtained from his product and the diagonal map $\pi \to \pi \times \pi$. If π is finite, then $R_0(\pi)$ is just the representation ring for π. Now, generalizing a construction due to Atiyah and Hirzebruch [9], for the case of compact groups he indicates how to construct a homomorphism $\alpha: R_i(\pi) \to K^{-i}(B\pi)$, where $B\pi$ is a classifying space for π. (In the context of Atiyah-Hirzebruch π is finite and one must construct a bundle over $B\pi$ given a representation of π. The construction for infinite π is nontrivial and a more detailed description is given by Mischenko [61].) Kasparov conjectures that the image of α is dense in the topology on $K^*(B\pi)$ induced by the filtration of $B\pi$ by compact sets. He claims to be able to prove this for discretely imbedded subgroups of $GL(n, C)$. Kasparov next outlines the construction of a homomorphism $\beta: K_i(B\pi) \to K^{-i}(C^*(\pi))$ dual to α. (It has been suggested to the author that the homomorphism β is closely related to the natural map from homology groups to surgery groups (cf. [36]).) By duality Kasparov claims that whenever α has dense range, it follows that β is a monomorphism. Kasparov's interest in β stems from its connection with the homotopy invariance of higher signatures for manifolds.

For M an oriented compact manifold of dimension $4k$, the *signature* of M, denoted $\mathrm{Sign}(M)$, is defined to be the signature (the number of positive eigenvalues minus the number of negative eigenvalues) of the symmetric nondegenerate bilinear form

$$H^{2k}(M, Q) \times H^{2k}(M, Q) \to Q$$

induced by the cup product. The Hirzebruch signature theorem [43] states that

$$\mathrm{Sign}\,(M) = \langle L(M), [M] \rangle ,$$

where $[M]$ is the fundamental class of M in $H_{4k}(M, \mathbb{Q})$, $L(M)$ is the L-class in $H^*(M, \mathbb{Q})$ and $\langle\ ,\ \rangle$ denotes the pairing of cohomology and homology. The L-class is the characteristic class obtained by pulling back the universal L-class in $H^*(BO, \mathbb{Q})$ defined by Hirzebruch, via the map from M to BO which classifies the tangent bundle of M, where BO is the classifying space for the orthogonal group. A number defined by evaluating a characteristic class at the fundamental class is called a *characteristic number*. Since the signature is obviously a homotopy invariant, it follows that this characteristic number is also; that is, if $\phi : M \to M'$ is a continuous orientation-preserving map from M to another compact oriented 4k-dimensional manifold M' which is a homotopy equivalence (but not necessarily a diffeomorphism, or even a smooth map) then

$$\langle L(M), [M] \rangle = \langle L(M'), [M'] \rangle .$$

Moreover, if M is simply connected, then it was shown by Browder [17] and Novikov [63] that this is the only characteristic number which is a homotopy invariant. However, if M is not simply connected then Novikov conjectured [64] that this is not the case.

Let M be an oriented compact n-dimensional manifold with $\pi_1(M) = \pi$ and let $f: M \to B\pi$ be a map defining an isomorphism on fundamental groups. If a is an element of $H^*(B\pi, \mathbb{Q})$, then $f^*(a)$ is in $H^*(M, \mathbb{Q})$ and

$$\langle L(M)\, f^*(a), [M] \rangle$$

is a characteristic number of M. Such numbers are called the *higher signatures* of M. One can show [61] using bordism groups that a homotopically invariant characteristic number must be a higher signature. Novikov conjectured [64] that the higher signatures for which a is the product of one-dimensional cohomology classes are homotopy invariant. This was verified by Rohlin [81], Farrell-Hsiang [35] and Kasparov [51]. Further, the homotopy invariance of the higher signatures corresponding to all cohomology classes has been established for many groups π. In fact, since no higher signatures are known which are not homotopy invariant, the generalized Novikov conjecture is that all higher signatures are homotopy invariant.

Using Poincaré duality $D: H^*(M,\bar{Q}) \to H_*(M,\bar{Q})$ we can rewrite

$$< L(M) f^*(a), [M] > = < a, f_*(DL(M)) > .$$

Then all higher signatures for M will be homotopy invariant if and only if $f_*(DL(M))$ in $H_*(M,Q)$ is homotopy invariant.

Now one can prove using the Atiyah-Singer index theorem or its generalization to K-homology that if $D^+ : \Omega^+ \to \Omega^-$ denotes the signature operator on M [10] and $[D^+]$ denotes the corresponding element in $K_0(M)$ defined using Atiyah's $\mathrm{Ell}(M)$, that $ch_*[D^+] = DL(M)$ and hence

$$f_*(DL(M)) = f_* ch_*[D^+] = ch_*(f_*[D^+]) .$$

Therefore, to establish the homotopy invariance of the higher signatures for M, it is sufficient to establish the homotopy invariance of $f_*[D^+]$. Kasparov claims to be able to prove that $(\beta \otimes 1)(f_*[D^+])$ is always homotopy invariant, where

$$\beta \otimes 1 : K_0(B\pi) \otimes Q \to K^0(C^*(\pi)) \otimes \bar{Q}$$

is the homomorphism alluded to earlier. Thus Kasparov reduces homotopy invariance of the higher signatures for M to verifying that $\beta \otimes 1$ is a monomorphism or that α has dense range. Hence he claims to be able to prove the homotopy invariance for the higher signatures for manifolds with fundamental group which can be discretely imbedded in $GL(n,C)$. The details have not yet appeared.

Mischenko and his colleagues have obtained [61] related results also by functional analytic techniques. Mischenko considers a Fredholm representation of a ring Λ with involution. This generalizes Atiyah's $\mathrm{Ell}(X)$

and consists of two symmetric representations ρ_1 and ρ_2 on $\mathfrak{L}(\mathfrak{H}_1)$ and $\mathfrak{L}(\mathfrak{H}_2)$ together with a Fredholm operator $F : \mathfrak{H}_1 \to \mathfrak{H}_2$ which satisfies $\rho_2(\lambda)F - F\rho_1(\lambda)$ is compact for λ in Λ. He introduces this to obtain a signature-type invariant for a free finitely generated Λ-module M equipped with a Hermitian form α. If $A = \{\lambda_{ij}\}$ is the matrix for α relative to some free basis for M, then one can define operators

$$A_{\rho_k} = \{\rho_k(\lambda_{ij})\}$$

acting on the direct sum $\mathfrak{H}_k^s$ of s copies of $\mathfrak{H}$ for $k = 1, 2$, where s is the rank of M. Further, we let $\tilde{F}$ denote the operator defined from $\tilde{\mathfrak{H}}_1$ to $\tilde{\mathfrak{H}}_2$ by the operator matrix with F on the diagonal. If we split the spaces $\tilde{\mathfrak{H}}_1 = \tilde{\mathfrak{H}}_1^+ \oplus \tilde{\mathfrak{H}}_1^-$ and $\tilde{\mathfrak{H}}_2 = \tilde{\mathfrak{H}}_2^+ \oplus \tilde{\mathfrak{H}}_2^-$ according to the positive and negative parts of $\tilde{A}_{\rho_1}$ and $\tilde{A}_{\rho_2}$, respectively, then the decomposition of $\tilde{F}$ into the matrix

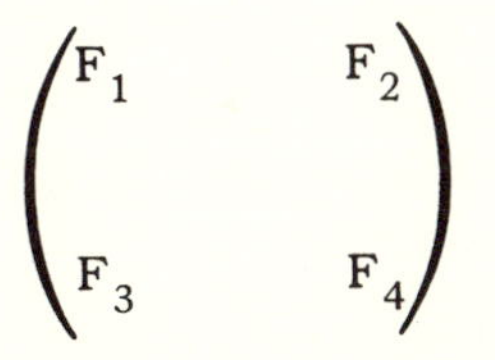

$$\begin{pmatrix} F_1 & F_2 \\ F_3 & F_4 \end{pmatrix}$$

has the property that F_2 and F_3 are compact and hence F_1 and F_4 are Fredholm operators. Finally, we define the signature invariant by $\mathrm{sign}_\rho(M, \alpha) = \mathrm{ind}(F_1) - \mathrm{ind}(F_4)$. This extends to define a homomorphism from $K_0^h(\Lambda)$ to Z, where $K_0^h(\Lambda)$ is the hermitian K-group defined by equivalence classes of finitely generated Λ-modules with a hermitian structure.

For $\Lambda = \ell^1(\pi)$, the group ring of a discrete group, Mischenko shows using a Fredholm representation ρ for Λ, how one can construct a bundle ξ_ρ over $B\pi$. (This is the same construction as that used by Kasparov to define his homomorphism α.) He also considers the problem of determining the subgroup $K^0(B\pi)$ generated by the bundles ξ_ρ arising from Fredholm representations of $\ell^1(\pi)$. In particular, he shows that if

$B\pi$ is homotopically equivalent to a compact Riemannian manifold with a metric of non-positive curvature in every two-dimensional direction, then this is all of $K(B\pi)$.

The connection of these results with the homotopy invariance of higher signatures is via a generalized multiply-connected Hirzebruch signature formula

$$\mathrm{sign}_\rho(\sigma(M)) = \langle L(M) f^* ch^* \xi_\rho, [M] \rangle ,$$

for ρ a Fredholm representation of $\ell^2(\pi)$, where $\sigma(M)$ is an element of $K_0^h(\ell^1(\pi))$ constructed from M. (This construction is not trivial and seems to be related to Kasparov's β.) Since the left hand side of the identity is homotopy invariant, then the homotopy invariance of the higher signatures defined by cohomology classes of the form $ch^*\xi_\rho$ follows. Therefore, Mischenko claims all higher signatures are homotopy invariant for manifolds with a fundamental group π for which $B\pi$ satisfies the condition on non-positive curvature. Additional results are claimed [61], for other classes of groups for which one can establish that cohomology classes of the form $ch^*\xi_\rho$ generate $H^*(B\pi, \bar{Q})$.

It would seem to this author that in basic outline of proof, the approaches of Kasparov and Mischenko to the homotopy invariance of higher signatures coincide. The point of view is different but very close. There would seem to be some possible advantage in Kasparov's approach if his homomorphisms α and β could be shown to exist more simply and possess good functorial properties. But we will have to wait for the promised details from Kasparov.

We conclude with brief comments on other results related to Ext. Firstly, it is possible to consider Ext based on either reals or the quaternions. One would either use real or quaternionic Hilbert space or in the case of R, use the complex case but carry along a conjugation mapping J satisfying $J^2 = -i$. Another generalization would be to the equivariant case in which a group acts. As might be expected finite groups cause little additional difficulty but compact groups are a different matter. Preliminary results in establishing an equivariant Ext-theory for compact groups have been obtained by Kaminker-Schochet [49] and Loebl-Schochet [56] and Schochet [86]. I should add that Kasparov's version of K-homology covers all these generalizations.

A different kind of generalization of Ext replaces the ideal of compact operators $\mathcal{K}$ by some other C^*-algebra. In [79] Pimsner, Popa, and Voiculescu replace $\mathcal{K}$ by $\mathcal{K} \otimes C(X)$ and obtain a two-variable functor, covariant in one and contravariant in the other. It is by exploiting both variables that they obtain the results already mentioned allowing the proof of homotopy invariance and the establishment of the short exact sequence for separable type I C^*-algebras. A more elaborate generalization of a similar nature has been recently announced by Kasparov [54] along with more precise statements of the relation of his work with ours. In another direction the von Neumann algebra which is a type I_∞ factor with unique closed ideal $\mathcal{K}(\mathfrak{H})$ can be replaced by a type II_∞ factor $\mathcal{O}$. Such an algebra also possesses a unique closed ideal $\mathcal{J}$. Whereas the dimension is a nonnegative integer valued function on the projections in $\mathcal{L}(\mathfrak{H})$, on $\mathcal{O}$ there is a real-valued dimension function and $\mathcal{J}$ is generated by the "finite dimensional" projections. Breuer showed [15], [16] that the notion of Fredholm operator could be generalized to this context and that the space of "generalized Fredholm operators" formed a classifying space for K-theory with real coefficients. He also showed that these K-groups could be defined using vector bundles which are "finite-dimensional" in the sense of a II_∞ factor. Analogously, the entire theory of extensions, at least for extensions of $\mathcal{J}$ by $C(X)$, can be carried out in this context. A proof of the analogue of the Weyl-von Neumann theorem was given by Zsido [104] (cf. [37], [26]). With that basically all the proofs in [22] carry over. This Ext defines K-homology with real coefficients, and hence is torsion-free, and can be paired with K-theory with real coefficients (cf. [26]).

As indicated in the first chapter, one of the original motivations for defining Ext was the classification problem for essentially normal operators. The matrix analogue of a normal operator is an n-normal operator. An operator T on $\mathfrak{H}$ is said to be n-normal if $\mathfrak{H}$ is isomorphic to the direct sum of n-copies of $\mathfrak{H}_0$ such that the matrix entries for T relative to this decomposition consist of commuting normal operators. The notion of essentially n-normal operator can also be defined and then classification modulo compact operators makes sense. The analogue of the spectrum for these operators is a certain compact subset of $M_n(C)$. Considerable results have been obtained by Pearcy-Salinas [71], [72], Salinas [83], [85], Paulsen [68], Paulsen-Salinas [69] and McGovern-Paulsen-Salinas [57] in solving the classification problem for essentially n-normal operators along the lines of that for the essentially normal case.

Now Ext presents possibilities as a tool in the study of C^*-algebras. As colorfully put by Effros, one can view the Ext theory for C^*-algebras as "noncommutative algebraic topology". In particular, Ext has been used by Z'ep [103], Green [40], and Rosenberg [82] to characterize the group C^*-algebra for several infinite families of solvable locally compact groups. For example, Z'ep shows that if we consider the group C^*-algebra for the group of all affine transformations on the real line, then it is an extension of $\mathcal{K}$ by $C(X)$, where X is the one-point compactification of two parallel lines. Since such extensions are classified by $Z \oplus Z$, he then determines which one he has by calculating two indices. It corresponds to $1 \oplus 1$.

Also one can use Ext as an isomorphism invariant on the AF algebras of Bratelli [14]. The closure $\mathfrak{A}$ of an increasing union $\bigcup_i \mathfrak{A}_i$ of finite dimensional C^*-subalgebras of $\mathfrak{L}(\mathfrak{H})$ containing the identity is called an AF-algebra. In case $\mathfrak{A}_i \cong \mathfrak{L}(C^{n_i})$, then $\mathfrak{A}$ is a UHF algebra. It is clear that n_i divides n_{i+1}; moreover, UHF algebras for which the groups $G = \lim Z/n_i Z$ are equal, are in fact isomorphic. Glimm showed [39] that the converse is also true. Pimsner and Popa [77] show that for a UHF algebra $\mathfrak{A}$, one has $\mathrm{Ext}^{str}(\mathfrak{A}) \cong G$ which also proves the converse. Another way of obtaining G was also given by Singer [90], and Smith-Smith [91] who showed that it is isomorphic to $\pi_1(U(\mathfrak{A}))$, where $U(\mathfrak{A})$ denotes the group of unitary elements in $\mathfrak{A}$. Pimsner and Popa [77], [76] also determine $\mathrm{Ext}^{\mathrm{str}}(\mathfrak{A})$ for $\mathfrak{A}$ an AF-algebra as $\mathrm{ext}^1_Z(\pi_1(U(\mathfrak{A})/_T), Z)$. This is done by extending the arguments indicated in Chapter 5 on the kernel of the inverse limit homomorphism to AF-algebras. Many of these results were also obtained by L. G. Brown.

Additional results have been obtained by Phillips and Raeburn [75] on extensions of $\mathcal{K}$ by AF-algebras. They show that if τ_1 and τ_2 are *-homomorphisms of an AF-algebra $\mathcal{C}$ into $\mathfrak{Q}(\mathfrak{H})$ such that τ_1 is trivial and $\|\tau_2 - \tau_1\|$ is sufficiently small, then τ_2 is also trivial. Also, they show that a homotopy (continuous in the point-norm topology) of *-homomorphisms $\pi_t\colon \mathcal{C} \to \mathfrak{Q}(\mathfrak{H})$ such that $\tau_0 = \pi \circ \sigma_0$ is trivial can be lifted to a homotopy of *-homomorphisms $\sigma_t : \mathcal{C} \to \mathfrak{L}(\mathfrak{H})$ such that $\tau_t = \pi \circ \sigma_t$. Additional results have been obtained by Phillips in [73] where he solves the homotopy lifting and perturbation problem for AF-algebras. Finally he solves the perturbation of C*-algebras problem for AF-algebras in [74].

Lastly, Ext has been calculated for some C*-algebras recently introduced by Cuntz [29]. The Cuntz C*-algebra O(n) is that generated by n isometries $\{S_i\}_{i=1}^n$, $n = 2, 3, \cdots \infty$ which satisfy $\sum_{i=1}^{n} S_i S_i^* = I$ for n finite or $\sum_{i=1}^{\infty} S_i S_i^* \leq I$ for n infinite. Cuntz showed that O(n) is independent of the choice of isometries. Pimsner and Popa showed [78] that $\mathrm{Ext}(O(n)) \cong Z_{(n-1)}$ for n finite and $\mathrm{Ext}(O(\infty)) = \{0\}$. Similar results were also obtained by L. G. Brown and by Paschke and Salinas [67] for $O(n) \otimes M_k$. These results enable many non-isomorphism results to be proved for the class $O(n) \otimes M_k$.

It seems clear that Ext will provide an increasingly effective tool for deciding such questions.

REFERENCES

[1] T. B. Andersen, Linear extensions, projections and split faces, J. Functional Anal. 17 (1974), 161-173.

[2] D. W. Anderson, Universal coefficient theorems for K-theory, to appear.

[3] J. Anderson, A C^*-algebra $\mathfrak{A}$ for which $\mathrm{Ext}(\mathfrak{A})$ is not a group, Ann. Math. 107 (1978), 455-458.

[4] W. B. Arveson, A note on essentially normal operators, Proc. Roy. Irish Acad. 74 (1974), 143-146.

[5] ________, Notes on extensions of C^*-algebras, Duke Math. J. 44 (1977), 329-355.

[6] M. F. Atiyah, K-theory, Lecture Notes by D. W. Anderson, W. A. Benjamin, Inc., New York-Amsterdam, 1967.

[7] ________, "Global Theory of Elliptic Operators," Proc. Internat. Conf. on Functional Analysis and Related Topics, Univ. of Tokyo Press, Tokyo, 1970.

[8] ________, "A survey of K-theory," Procedures of the Conference on K-theory and Operator algebras, Athens, Georgia; Springer-Verlag, Lecture Note Series, 575, 1-9, 1977.

[9] M. F. Atiyah and F. Hirzebruch, Vector bundles and homogeneous spaces, Proc. Sympos. Pure Math., vol. 3, Amer. Math. Soc., Providence, R. I., 1961, 7-38.

[10] M. F. Atiyah and I. M. Singer, The index of elliptic operators I, III, Ann. Math. (2) 87 (1968), 484-530, 546-604.

[11] P. Baum, K-homology, Class Notes, Brown U., 1977.

[12] I. D. Berg, An extension of the Weyl-von Neumann theorem to normal operators, Trans. Amer. Math. Soc. 160 (1971), 365-371.

[13] L. Boutet de Monvel, On the index of Toeplitz operators of several complex variables, Inventiones Math. 50 (1979), 249-272.

[14] O. Bratteli, Inductive limits of finite dimensional C^*-algebras, Trans. Amer. Math. Soc. 171 (1972), 195-234.

[15] M. Breuer, Fredholm theories on von Neumann algebras I, II, Math. Ann. 178 (1968), 243-254; 180 (1969), 313-325.

[16] M. Breuer, Theory of Fredholm operators and vector bundles relative to a von Neumann algebra, Rocky Mountain J. Math. 3 (1973), 383-429.

[17] W. Browder, Homotopy type of differential manifolds, Coll. Algebraic topology, August 1962, Aarhus Univ., 1-10.

[18] L. G. Brown, Operator algebras and algebraic K-theory, Bull. Amer. Math. Soc. 81 (1975), 1119-1121.

[19] ———, Extensions and the structure of C^*-algebras, Convegno Teoria degli operatori indice e teorra K. (Rome, Ottobre 1975) Symposia Mathematica XX, Rome, 1977.

[20] L. G. Brown, R. G. Douglas, and P. A. Fillmore, Unitary equivalence modulo the compact operators and extensions of C^*-algebras, Proc. Conf. on Operator Theory, Springer-Verlag Lecture Notes 345, 58-128, 1973.

[21] ———, Extensions of C^*-algebras, operators with compact self-commutators, and K-homology, Bull. Amer. Math. Soc. 79 (1973), 973-978.

[22] ———, Extensions of C^*-algebras and K-homology, Ann. Math (2) 105 (1977), 265-324.

[23] L. G. Brown and C. Schochet, K_1 of the compact operators is zero, Proc. Amer. Math. Soc. 59 (1976), 119-122.

[24] F. P. Cass and V. P. Snaith, On C^*-algebra extensions of Lie groups, to appear.

[25] S. J. Cho, Strong extensions vs. weak extensions of C^*-algebras, Canad. Math. Bull. 21 (1978), 143-147.

[26] ———, Extensions relative to a II_∞ factor, Proc. Amer. Math. Soc. 74 (1979), 109-112.

[27] M. D. Choi and E. G. Effros, The completely positive lifting problem, Ann. Math. 104 (1976), 585-609.

[28] L. A. Coburn, The C^*-algebra generated by an isometry, II, Trans. Amer. Math. Soc. 137 (1969), 211-217.

[29] J. Cuntz, Simple C^*-algebras generated by isometries, Comm. Math. Phys. 57 (1977), 177-185.

[30] A. M. Davie, Classification of essentially normal operators, Spaces of Analytic Functions, Springer-Verlag Lecture Notes 512, 31-55, 1976.

[31] J. Dixmier and A. Douady, Champs continus d'espaces hilbertiens, Bull. Soc. Math. de France 91 (1963), 227-283.

[32] R. G. Douglas, Banach algebra techniques in operator theory, Academic Press, New York, 1972.

[33] R. G. Douglas, The relation of Ext to K-theory, Convegno Teoria degli operatori indice e teorra K. (Rome, Ottobre 1975) Symposia Matematica XX, Rome, 1977.

[34] S. Eilenberg, and N. Steenrod, Foundations of algebraic topology, Princeton, 1952.

[35] F. T. Farrell and W. C. Hsiang, A geometric interpretation of the Künneth formula for algebraic K-theory, Bull. Amer. Math. Soc. 74 (1968), 548-553.

[36] ________, On the rational homotopy groups of the diffeomorphism groups of discs, spaces and aspherical manifolds, Proc. Symposia Pure Math. 32 (1978), 325-337.

[37] P. A. Fillmore, Extensions relative to semi-finite factors, Convegno Teoria degli operatori indice e teorra K. (Rome, Ottobre 1975) Symposia Mathematica XX, Rome, 1977.

[38] P. A. Fillmore, J. G. Stampfli and J. P. Williams, On the essential numerical range, the essential spectrum, and a problem of Halmos, Acta Sci. Math. (Szeged) 33 (1972), 179-192.

[39] J. G. Glimm, On a certain class of operator algebras, Trans. Amer. Math. Soc. 95 (1960), 318-340.

[40] P. Green, C^*-algebras of transformation groups with smooth orbit space, Pacific J. Math. 72 (1977), 71-97.

[41] P. R. Halmos, Ten problems in Hilbert space, Bull. Amer. Math. Soc. 76 (1970), 887-933.

[42] J. W. Helton and R. Howe, Traces of commutators of integral operators, Acta Math. 136 (1976), 271-305.

[43] F. Hirzebruch, Topological methods in algebraic geometry, Springer-Verlag, New York, 1966.

[44] L. Hodgkin, On the K-theory of Lie groups, Topology 6 (1967), 1-36.

[45] D. Husemoller, Fiber Bundles, McGraw-Hill, New York, 1966.

[46] D. S. Kahn, J. Kaminker, and C. Schochet, Generalized homology theories on compact metric spaces, Mich. Math. J. 24 (1977), 203-224.

[47] J. Kaminker and C. Schochet, Steenrod homology and operator algebras, Bull. Amer. Math. Soc. 81 (1975), 431-434.

[48] ________, K-theory and Steenrod homology: Applications to the Brown-Douglas-Fillmore theory of operator algebras, Trans. Amer. Math. Soc. 227 (1977), 63-107.

[49] ________, Analytic equivariant K-homology, Geometric Applications of Homotopy Theory I, Springer-Verlag, Lecture Notes 657, 1978.

[50] M. Karoubi, K-theory, An Introduction, Springer-Verlag, Berlin-Heidelberg, New York, 1978.

[51] G. G. Kasparov, On the homotopy invariance of the rational Pontryagin numbers, Dokl. Akad. Nauk SSSR 190 (1970), 1022-1025.

[52] ———, The generalized index of elliptic operators, Funktional Anal. i Prilozen 7 (1973), 82-83.

[53] ———, Topological invariants of elliptic operators I: K-homology, Math. USSR Izvestija 9 (1975), 751-792.

[54] ———, K-functor in the extension theory of C^*-algebras, Funktional. Anal. i Prilozen (to appear).

[55] P. Kohn, A homotopy theory of C^*-algebras, thesis, University of Penn., 1972.

[56] R. Loebl and C. Schochet, Covariant representations on the Calkin algebra I, Duke Math. J. 45 (1978), 721-734.

[57] R. McGovern, V. Paulsen and N. Salinas, A classification theorem for essentially binormal operators, to appear.

[58] J. Milnor, On the Steenrod homology theory, Berkeley, 1961 (mimeo).

[59] ———, Introduction to Algebraic K-theory, Ann. of Math. Studies 72, Princeton, N. J., Princeton Univ. Press, 1971.

[60] J. Milnor and J. Stasheff, Lecture Notes on characteristic classes, Ann. of Math. Studies 76, Princeton, N. J., Princeton Univ. Press, 1974.

[61] A. S. Mischenko, Hermitian K-theory. The theory of characteristic classes and methods of functional analysis, Russian Math. Surveys 31 (2) (1976), 71-138.

[62] J. von Neumann, Charakteristerung des Spectrums eines Integral Operators, Hermann, Paris, 1935.

[63] S. P. Novikov, Homotopically equivalent smooth manifolds I, Iz. Akad. Nauk SSR Ser. Mat. 28 (1964), 365-474.

[64] ———, Homotopic and topological invariance of certain rational classes of Pontrjagin, Dokl. Akad. Nauk SSR 162 (1965), 1248-1251.

[65] D. P. O'Donovan, Quasidiagonality in the Brown-Douglas-Fillmore theory, Duke Math. J. 44 (1977), 767-776.

[66] C. L. Olsen and W. R. Zame, Some C^*-algebras with a single generator, Trans. Amer. Math. Soc. 215 (1976), 205-217.

[67] W. L. Paschke and N. Salinas, Matrix algebras over O_n, Mich. Math. J., 26 (1979), 3-12.

[68] V. Paulsen, Weak compalence invariants for essentially n-normal operators, Amer. J. Math. 101 (1979), 979-1006.

[69] V. Paulsen and N. Salinas, Two examples of non-trivial essentially n-normal operators, Indiana Univ. Math. J. 28 (1979), 711-724.

[70] C. Pearcy and N. Salinas, Operators with compact self-commutator, Canad. J. Math. 26 (1974), 115-120.

[71] ________, The reducing essential matricial spectrum of an operator, Duke Math. J. 42 (1975), 423-434.

[72] ________, "Extensions of C^*-algebras and the reducing essential matricial spectra of an operator," Proceedings of the conference on K-theory and Operator algebras, Athens, Georgia; Springer-Verlag Lecture Note Series 575, 95-112, 1973.

[73] J. Phillips, On extensions of AF-algebras, Amer. J. Math. 101 (1979), 957-968.

[74] ________, Perturbations of AF-algebras, Can. Math. J. 31 (1979), 1013-1016.

[75] J. Phillips and I. Raeburn, On extensions of AF-algebras, Amer. J. Math. 101 (1979), 957-968.

[76] M. Pimsner, On the Ext-group of an AF-algebra, II, Rev. Roum. Math. Pures et Appl. 24 (1979), 1085-1088.

[77] M. Pimsner and S. Popa, On the Ext-group of an AF-algebra, Rev. Roum. Math. Pures Appl. 23 (1978), 251-267.

[78] ________, The Ext-groups of some C^*-algebras considered by J. Cuntz, Rev. Roum. Math. Pures Appl. 23 (1978), 1069-1076.

[79] M. Pimsner, S. Popa, and D. Voiculescu, Homogeneous C^*-extensions of $C(X) \otimes \mathcal{K}(\mathfrak{H})$. Part I, J. of Operator Theory, 1 (1979), 55-108.

[80] I. Raeburn, K-theory and K-homology relative to a II_∞-factor, Proc. Amer. Math. Soc. 71 (1978), 294-298.

[81] V. A. Rohlin, The Pontryagin-Hirzebruch class of codimension 2, Izv. Akad. Nauk SSR Ser. Mat. 30 (1960), 705-718.

[82] J. Rosenberg, The C^*-algebras of some real and p-adic solvable groups, Pacific J. Math. 65 (1976), 175-192.

[83] N. Salinas, Extensions of C^*-algebras and essentially n-normal operators, Bull. Amer. Math. Soc. 82 (1976), 143-146.

[84] ________, Homotopy invariance of Ext( ), Duke Math. J. 44 (1977), 77-94.

[85] ________, Hypoconvexity and essentially n-normal operators, Trans. Amer. Math. Soc. 256 (1979), 325-351.

[86] C. Schochet, Covariant representations on the Calkin algebra II, (preprint).

[87] G. Segal, "K-homology theory and algebraic K-theory," Proceedings of a Conference on K-theory and Operator algebras, Athens, Georgia: Springer-Verlag Lecture Notes Series 575, 1977.

[88] I. M. Singer, Future extensions of index theory and elliptic operators, Prospects in Mathematics, Ann. Math. Studies 70, Princeton (1971), 171-185.

[89] I. M. Singer, "Some remarks on operator theory and index theory," Proceedings of a Conference on K-theory and Operator algebras, Athens, Georgia; Springer-Verlag Lecture Notes Series 575, 128-138, 1977.

[90] ________, On the classification of UHF-algebras, (unpublished).

[91] M. B. Smith and L. Smith, On the classification of UHF C^*-algebras, (unpublished).

[92] E. H. Spanier, Function spaces and duality, Ann. Math. (2) 70 (1959), 338-378.

[93] E. H. Spanier and J. H. C. Whitehead, Duality in homotopy theory, Mathematika 2 (1955), 56-80.

[94] N. Steenrod, Regular cycles of compact metric spaces, Ann. Math. (2) 41 (1940), 883-851.

[95] D. Sullivan, Geometric topology seminar notes, Princeton 1970.

[96] J. L. Taylor, Banach algebras and topology, Algebras in Analysis, Academic Press, London (1975), 118-186.

[97] U. Venugopalkrishna, Fredholm operators associated with strongly pseudo convex domains in C^n, J. Functional Anal. 9 (1972), 349-372.

[98] J. Vesterstrøm, Positive linear extension operators for spaces of affine functions, Israel J. Math. 16 (1973), 203-211.

[99] D. Voiculescu, A non-commutative Weyl-von Neumann theorem, Rev. Roum, Math. Pures et Appl. 21 (1976), 97-113.

[100] ________, On a theorem of M. D. Choi and E. G. Effros, Preprint.

[101] H. Weyl, Über beschrankte quadratische Formen deren Differenz vollstetig ist, Rend. Circ. Mat. Palermo 27 (1909), 373-392.

[102] G. W. Whitehead, Generalized homology theories, Trans. Amer. Math. Soc. 102 (1962), 227-283.

[103] D. N. Z'ep, On the structure of the group C^*-algebra of the group of affine transformations of the line (Russian), Funktional. Anal. i Prilozen 9 (1975), 63-64.

[104] L. Zsido, The Weyl-von Neumann theorem in semifinite factors, J. Functional Anal. 18 (1975), 60-72.

INDEX

INDEX OF SYMBOLS

Library of Congress Cataloging in Publication Data

Douglas, Ronald G
C*-algebra extensions and K-homology.

(Hermann Weyl lectures; 1978)
Bibliography: p.
Includes index.
1. C*-algebras. 2. K-theory. 3. Algebra, Homological. I. Title. II. Series.
QA326.D68 512′.55 80-424
ISBN 0-691-08265-0
ISBN 0-691-08266-9 pbk.